HACKING DIVERSITY

PRINCETON STUDIES IN

CULTURE AND TECHNOLOGY

Princeton Studies in Culture and Technology

Tom Boellstorff and Bill Maurer, series editors

This series presents innovative work that extends classic ethnographic methods and questions into areas of pressing interest in technology and economics. It explores the varied ways new technologies combine with older technologies and cultural understandings to shape novel forms of subjectivity, embodiment, knowledge, place, and community. By doing so, the series demonstrates the relevance of anthropological inquiry to emerging forms of digital culture in the broadest sense.

Hacking Diversity: The Politics of Inclusion in Open Technology Cultures by Christina Dunbar-Hester

Hydropolitics: The Itaipú Dam, Sovereignty, and the Engineering of Modern South America by Christine Folch

The Future of Immortality: Remaking Life and Death in Contemporary Russia by Anya Bernstein

Chasing Innovation: Making Entrepreneurial Citizens in Modern India by Lilly Irani

Watch Me Play: Twitch and the Rise of Game Live Streaming by T. L. Taylor

Biomedical Odysseys: Fetal Cell Experiments from Cyberspace to China by Priscilla Song

Disruptive Fixation: School Reform and the Pitfalls of Techno-Idealism by Christo Sims

Everyday Sectarianism in Urban Lebanon: Infrastructures, Public Services, and Power by Joanne Randa Nucho

Democracy's Infrastructure: Techno-Politics and Protest after Apartheid by Antina von Schnitzler

Digital Keywords: A Vocabulary of Information Society and Culture edited by Benjamin Peters

Sounding the Limits of Life: Essays in the Anthropology of Biology and Beyond by Stefan Helmreich with contributions from Sophia Roosth and Michele Friedner

Hacking Diversity

The Politics of Inclusion in Open Technology Cultures

Christina Dunbar-Hester

PRINCETON UNIVERSITY PRESS
PRINCETON AND OXFORD

Published by Princeton University Press
41 William Street, Princeton, New Jersey 08540
6 Oxford Street, Woodstock, Oxfordshire OX20 1TR

press.princeton.edu

Library of Congress Cataloging-in-Publication Data

Names: Dunbar-Hester, Christina, 1976– author.
Title: Hacking diversity : the politics of inclusion in open technology cultures / by Christina Dunbar-Hester.
Description: Princeton, New Jersey : Princeton University Press, [2020] | Series: Princeton studies in culture and technology | Includes bibliographical references and index.
Identifiers: LCCN 2019019271| ISBN 9780691182070 (hardcover : acid-free paper) | ISBN 9780691192888 (pbk. : acid-free paper)
Subjects: LCSH: Hacktivism. | Computers and women. | Open source software—Social aspects. | Multiculturalism.
Classification: LCC HV6773 .D855 2020 | DDC 303.48/34—dc23
LC record available at https://lccn.loc.gov/2019019271

British Library Cataloging-in-Publication Data is available

Editorial: Fred Appel and Jenny Tan
Production Editorial: Natalie Baan
Production: Erin Suydam
Publicity: Nathalie Levine and Kathryn Stevens

This book has been composed in Adobe Text and Gotham

Printed on acid-free paper. ∞

Printed in the United States of America

10 9 8 7 6 5 4 3 2 1

CONTENTS

ILLUSTRATIONS

Figures

Table

ACKNOWLEDGMENTS

This book has progressed from "scattered thoughts" to "idea" to "research" to "writing" and now to "written," but I'm sort of at a loss to explain how. I am quite sure I owe a lot to others. I am surrounded by communities of people engaging with our worlds, whose reflections and actions generate my own engagement and points of connection. The people whose activities are the basis for analysis in this book are due the most prominent recognition. My presence—as a stranger and "analyst" whose presence (let alone utility) in their midst was not immediately legible—has been patiently and trustingly welcomed by a host of people, online and off. They generously offered me their time and candor, and sometimes opened up painful memories; I hope I have been able to do justice to their stories and efforts in these pages. It would be improper to name them, given ethnographic convention and the level of personal and professional sensitivity of some of this material, but I offer them my profoundest thanks.

In addition, I am humbled by the generosity and dazzling expertise of a company of scholars and friends without whom my work would literally not be possible (to say nothing of tolerable, let alone ever enjoyable). Biella Coleman has been an invaluable interlocutor since the beginning of this project, providing feedback both shrewd and gracious at critical moments, as has Chris Kelty. Laura Portwood-Stacer, always a generous and thoughtful sounding board, provided early listening and guidance throughout. I am especially indebted to constructive readers of chapter drafts and earlier iterations of this work: Steph Alarcón, Chris Anderson, Ron Eglash, Lucas Graves, Silvia Lindtner, Mike Palm, Luke Stark, and Sarah Myers West. The book's analysis was further shaped through focused conversation, literature recommendations, stray comments, and everything in between from Mike Ananny, Sarah Banet-Weiser, Jonah Bossewitch, Jack Bratich, Paula Chakravartty, A. J. Christian, Kristina Clair, Chris Csíkszentmihályi, Nathan Ensmenger, Megan Finn, Laura Forlano, Ellen Foster, Mel Gregg, Seda Gürses, DeeDee Halleck, Randy Illum, Natalie Jeremijenko, Dorothy Kidd, Ron Kline, Daniel Kreiss,

Chenjerai Kumanyika, Deepa Kumar, Javier Lezaun, Manjari Mahajan, Charlton McIlwain, Mara Mills, Cyrus Mody, Lilly Nguyen, Wazhmah Osman, Victor Pickard, Joseph Reagle, Bryce Renninger, Carrie Rentschler, Alessandra Renzi, Adrienne Shaw, Christo Sims, Johan Söderberg, Jonathan Sterne, Tom Streeter, Sophie Toupin, Lee Vinsel, Cristina Visperas, Judy Wajcman, and Todd Wolfson. Dubious credit (but sincere gratitude) for the pun in the title of the book's conclusion goes to Nick Seaver. I also thank my undergraduate and graduate students at the University of Southern California and Rutgers University who offered me the opportunity to work out some of my thoughts in conversation with them and shared theirs in return.

Fred Appel has been an engaged, rigorous, and enthusiastic editor. It has been a sincere pleasure to work with him. I am also profoundly grateful to my anonymous reviewers at Princeton University Press, and to the PUP staff, especially Jenny Tan, Natalie Baan, and Thalia Leaf, for their stellar assistance in bringing the book into being; Sarah McIntosh was an invaluable indexer. Research in progress has been presented at scholarly venues along the way, and I thank various audiences for productively critical feedback, including the University of Washington Information School, Oslo Metropolitan University's Work Research Institute, UCLA Information Studies Department, USC Digital Humanities Program, and the Annenberg School at the University of Pennsylvania. Josh Kun and Sarah Banet-Weiser at USC Annenberg helped me carve out time to work on this book at key stages in its development and completion. Crack research assistance was provided by Soledad Altrudi (who cheerfully logged several semesters on this project), as well as Hyun-Tae (Calvin) Kim, and Lin Zhang (and Fanny Ramirez at Rutgers). Also at USC, Annenberg colleagues showed kindness (and lent bicycles, may they rest in peace) as I was attempting to adjust to California, so thank you, Jonathan Aronson, François Bar, Willow Bay, Manuel Castells, Taj Frazier, Hernan Galperin, Larry Gross, Andrea Hollingshead, Henry Jenkins, Josh Kun, Carmen Lee, Mark Lloyd, Peggy McLaughlin, Lynn Miller, Gordon Stables, Alison Trope, Miki Turner, Dmitri Williams, and Ernie Wilson. Juan De Lara and Andy Lakoff have been delightful coconspirators. Special thanks to Hye Jin Lee for invitations to step off campus. ASCJ staff Ally Arguello, Jordan Gary, Sarah Holterman, Christine Lloreda, Raymond Marquez, Frank Miuccio, David Mora, and Billie Shotlow provided lots of help streamlining administrative matters large and small. At my former workplace, a campus exhibition called "Trans Technology: Circuits of Culture, Self, and Belonging" that I curated with Bryce Renninger in 2012–13 fed into this research in a few nonlinear ways, so I thank Bryce, Rutgers's Institute for Women and Art, and supportive former colleagues in Journalism & Media Studies and Women's and Gender Studies for that detour.

This project was seeded with substantial support from the National Science Foundation for research travel and time (Science, Technology and Society, award 1026818, coawarded to Gabriella Coleman). Kelly Joyce was a supremely professional and supportive program officer at the NSF, helping me find a way to use this award during a transition in my academic career. Staff at New York University helped administer this grant, no small feat with both principal investigators moving in and out of NYU during the period of the grant; thanks to John T. Johnson in the Office of Research for steering us through the rocks. Any opinions, findings, and conclusions or recommendations expressed in this material are mine, and do not necessarily reflect the views of the National Science Foundation. The Virtual Knowledge Studio in Amsterdam hosted me as a fellow in 2010 in very early stages of this project.

Some other people were here with me too (even when we were in different time zones, which was quite often, and then oftener, sigh). Comrades who fed, watered, sheltered, and entertained me during these years include Tucker and Teresa Aaron, Chris Anderson, Peter Asaro and Diana Mincyte, José Chan, John Cheney-Lippold, Steve Chong and Kiran Gajwani, David Copenhafer, Robert Davis, Marya Doerfel, Ingrid Erickson, Cassie Fennell and Dan Topor, Pete Fried, Jayson Harsin, Vicky Karkov, Jessica Kaufman, Jason Livingston, Marie Leger, Khaled Malas, Cynthia Mason, Joe Mendelson, ICP, Rüth Rosenberg, Willy Schofield and Ian Lay, Set Sokol, Felix Teitelbaum, Dan Thalhuber, M.W. Tidwell, Neerja Vasishta, and Brad "Salvatore" Waskewich. My aunt and uncle, Kathy and Roger Later, helped soften my LA landing midway into this project. I have appreciated the sunny quiet of the Hoversten household, where much of this book was drafted. My parents, Elaine Dunbar and Al Hester, supplied the bedrock for a lifetime of reading, writing, and social inquiry. At the risk of redundancy, I thank again several of the intellectual compadres already referenced, this time for their "pure" (if occasionally polluting) love and friendship: Victor, Lucas, Laura, Other Laura, Mike, Other Mike, Wazhmah, Steph, Cristina, Todd, Bryce, Christo, Deepa, Jack, Jay, SBW. Boundless gratitude is owed to Dug and Noodles for warmth, comfort, and cheering. And to Anna for always being on hand with warm socks, cookies, and general sisterly support. And to Erik, for being by my side as we face our uncertain future.

HACKING DIVERSITY

1

Introduction

> Why did this come up as an issue when it did? Well . . . People started tracking instances of overt sexism and talking about it. The default [had been], sigh, "another one of those [bad] things happened to me." Then there were the "groping is happening at your events" [discussions], sometimes people were blogging about this . . . Women who had been in the community long enough got to the point [where] they said, it's ok to have this conversation [now]. It's critical that at entry-level stuff you—*we*—don't fuck this stuff up [because then we will turn people off and lose them forever].[1]
>
> —INTERVIEWEE, MARCH 2012

A reader might reasonably assume that the opening quote refers to recent revelations of sexual harassment (and worse) in Hollywood. Starting in the fall of 2017, a series of accusations were leveled against prominent men in the entertainment industry, starting with film producer Harvey Weinstein. This sparked a wave of "#MeToo" public sphere discussions of harassment, misconduct, and unequal treatment of women that circulated in the media, around water coolers, on social media platforms, and in schools and homes.[2]

1. Interview, Meg, March 2012, Boston, MA.

2. Initially conceived by civil rights activist Tarana Burke in 2006, "#MeToo" drew women's individual experiences with sexual harassment into collectivity, creating space for empathy, solidarity, and activism. In the wake of Trump's election to the presidency, there was renewed urgency for attention to these issues. See Gibson et al. 2019; Ohlheiser 2017.

In a very short span of time, multiple powerful, high-profile men in politics, entertainment, and a few other fields were accused of wrongdoing, and in many cases terminated from their jobs. As of 2019, it is clear that this cultural moment is not over, and its final outcomes have yet to be written.

But this quote is from an interview I conducted in 2012 and pertains to a different cultural field entirely—the realm of *open technology*, where makers, programmers, and hackers mingle (online and off) to circulate knowledge and cultural goods, share technical problems and solutions, and generally revel in exploring technical puzzles related to coding and to craft. The interviewee was describing a dominant trend in these circles: snowballing advocacy around diversity and inclusion in voluntaristic technical communities, which had become pronounced as early as 2006. As the #MeToo momentum makes evident, this advocacy in open-technology cultures is an instantiation of a much wider social phenomenon, a collective reckoning with gendered imbalances in social power. This being said, there are features that make open-technology cultures distinct; they are largely convened in places where human resources departments or equal opportunity legislation do not hold sway, as they are voluntaristic, and they are (or have traditionally been) governed relatively informally. This means that participants have historically had little formal recourse to redress instances of either abuse or subtler unequal standing. Moreover, specific cultural barriers to addressing these issues, including the belief that these communities are open to whoever "wants to be there," have tended to perpetuate the notion that if some people are not there, it is because they do not wish to be. Lastly, though this quote makes reference to ugly behavior ("groping"), it would be a mistake to think that advocacy in these communities primarily pertains to ending violent mistreatment. Much more of it has to do with more mundane yet still laborious efforts to place inclusion at the fore and build up infrastructures for support.

Advocacy around "diversity" in software and hackerspace[3] communities, which spans the past decade and continues today, gathered momentum after the 2006 release of a European Union policy study that indicated that fewer than 2 percent of participants in free/libre and open-source software (FLOSS)[4] were women.[5] This appeared to set FLOSS apart from proprietary software development; while still heavily masculine, women's rate of

3. Hackerspaces are community workspaces where people with interest in computers, craft, and other types of fabrication come together to socialize and collaborate.

4. "Free" and "libre" are interchangeable (though the latter nods to languages that are not English), while "open" means something different (see chapter 2).

5. Nafus et al. 2006.

participation in non-FLOSS development was around 28 percent.[6] Participants were galvanized by these findings about FLOSS—what about their communities could account for such a dramatically low rate of participation by women?[7] What interventions were appropriate and effective to change this? A subsequent study, administered in 2013, showed that FLOSS participants who identified as nonmale had climbed to approximately 14 percent.[8] This shows that *something* was happening within these communities, resulting in an increase in participation by women and other people who did not identify as men.

Interventions to change the constitution of and practices within open-technology cultures should be understood as social analysis. Diversity advocacy flows from the same impulse to remake the world that FLOSS does. It is best understood as the practices that result from practitioners' shared belief that there is an error in how open-technology communities have been constituted, one that will leave a corresponding negative imprint on their technical output, which is how they intend to shape the world. Thus their impulse is to correct the "error" in the community (and not to reevaluate whether technology is the seat of progress). In attending to the processes and practices of diversity advocacy, we can observe how this advocacy flows from the social world of FLOSS itself, and attempts to remake it, and thus the world.

At the heart of this book lies the question: What happens when ordinary people try to define and tackle a large social problem? Though open-technology communities possess features different from the culture at large, they nevertheless constitute a laboratory for the voluntaristic address of social inequality. One special feature is, of course, their orientation around technology, which means they are beholden to the cultural legacies of computing and engineering in important ways. Another is their level of commitment to self-governance and autonomy. The hacker ethic includes a devotion to hands-on problem solving, which, this book argues, has led to open-technology enthusiasts trying to hack their communities in real time. In other words, some have addressed their communities with an approach

6. Nafus et al. 2006: 4.

7. It is worth pointing out that industry data may overstate the presence of women in technical fields due to conflation of women employed *overall* in these industries versus women employed in *technical* positions. So it is possible that the glaring 2006 statistics about FLOSS overstated how special FLOSS was in this regard. Thanks to Chris Kelty for discussion on this point.

8. Callahan et al. 2016: 575. Note that the framing of "nonmale" gender shifted during this period from "women" to "women and other people who did not identify as men."

that can be characterized as, *Hey, our culture is informal and constituted by shared interest in taking things apart and putting them back together again, so how hard could it be to change?*

Though this book is eminently sympathetic to these impulses, it shows is that there is a problem of scale for voluntaristic technologists hoping to reframe power relations. Part of the issue is that DIY interventions are insufficient to take on structural social problems (which is not a shortcoming of these communities, or their members' efforts). But it is also worth zeroing in on the analysis itself. Though "diversity in tech" discourse is emanating from many quarters in our current historical moment, it is important that the mandate of open-technology cultures is not identical to that of industry or of higher education. Here, the reasons for engagement with technology nominally include experiencing jouissance and a sense of agency. This is experienced through, yet not reducible to, community members' engagement with technology. If we tease apart the emancipatory politics from the technical engagement, we find that the calls for inclusion and for reframing power relations are not only about technical domains; rather, they are about agency, equity, and self-determination at individual and collective levels. This book argues that the social analysis offered by diversity advocates in open-technology cultures is important, but often incomplete. Calls for diversity in technical participation stop short of calls for justice—and are certainly not interchangeable. A consequence of this elision or slippage is that diversity work has the troubling potential to feed back into status quo arrangements of social and economic power that advocates are nominally critiquing. Therefore, a robust appraisal of power, and of technology's role in reproducing social orders, is required. This has implications that extend beyond open-technology communities. Careful attention to calls for inclusion and reconfiguring power in open-technology communities may, ironically, reach their fullest potential if they disentangle technology from agency. These two points—understanding how diversity advocacy does and does not scale, and where boundaries are drawn in technical communities' critical, reflexive attention to social and technical order—are this book's contributions to social analysis that can enhance and multiply diversity advocates' efforts to generate justice.

This book argues that this diversity advocacy in open-technology cultures holds the keys to a broader emancipatory politics, which is not *technological* emancipation. Precisely because their mandate is not, for example, to capture a wider market share or ready a national workforce, diversity advocates in open-technology communities have the space to articulate not only the potentials of technological engagement but also the limits. A rigorous analysis of power that emanates from technology-oriented communities is vital as conversations about power and inclusion roil science, technology, engineering, and math (STEM) industries and sites of education in our technologically advanced but socially and economically unequal societies. Ironically, *technical* inclusion may be a red herring when elemental equities have not been established.

The problem of scale can be addressed only through a more incisive analysis. It is ironic that the *hack*, the patch used to fix a bug, which had served practitioners well for technical problems in technical communities of a certain size, is not particularly well suited to the matters that concern diversity advocates. Entrenched social problems are hard to hack away. A more expansive critique that includes labor, class, and the transnational political economy of the material conditions that support Global North hacking is needed if these advocates wish to maximize the potentials of their emergent analysis. This is not a criticism of the individuals who toil at diversity advocacy. To the contrary, their activities represent sincere, vital, and caring[9] energies directed toward improving their communities and, it is hoped, the wider society. Diversity advocates in voluntaristic hacking settings are in a unique and influential position from which to launch a critique, as they are not beholden to institutions and not formally circumscribed by the power relations of workplaces. In other words, their power to effect structural change is limited, but their power to propagate social analysis of the stakes of diversity in tech is great. This is why it is important to bring their analysis into sharper focus.

This book uses the empirical site of advocacy around diversity in software and hackerspace communities to assess engagement with technology as a site of purposive political action. It explores multiple framings surrounding the overlapping issues of who participates in amateur technology cultures, to what ends, and with what consequences. My project here is distinctly *not* to ask (or answer) questions such as, "Why aren't there more women in STEM?" or "How can we bring more African American or Latinx people into STEM?" Rather, I uncover a range of motivations behind amateur interventions into diversity questions, in order to evaluate the political potentials and limitations of such projects. The multiple framings of who participates in technology development, and to what end, are taken as objects of inquiry in

9. As Maria Puig de la Bellacasa writes, care "signifies: an affective state, a material vital doing, and an ethico-political obligation" (2011: 90). See also Martin, Myers, and Viseu 2015.

their own right. The book argues that *technology* is a fraught concept to place in a central role in a project of emancipation, requiring special attention.

Give Me a Hackerspace and I Will Make the World

Hackers exhibit enthusiastic faith in their ability to effect change.[10] The sites of engagement with technology around which diversity advocacy is occurring can be grouped together under the umbrella of *open technology*, especially but not limited to free and open-source software. FLOSS is a set of practices for the distributed, collaborative creation of code that is made openly available through a reinterpretation of copyright law; it is also an ideologically charged mode of production and authorship that seeks to reorient power in light of participants' understandings of the moral and technical possibilities presented by the internet.[11] Hackerspaces are a cognate offline phenomenon, community workspaces where people with interest in computers, craft, and other types of fabrication come together to socialize and collaborate. These sites are far from monolithic, but they are more alike—bound together by a shared (if not singular) political and technical imaginary—than they are different. *Hacking* here is about an expression of agency, and not necessarily a desire to trespass or "own hard" (though some hacking subcultures possess this feature).[12] *Open technology* broadens the ethos of FLOSS to encompass software, hardware, and other cultural artifacts that proponents believe should be left open for the purposes of modification, reinterpretation, and refashioning toward purposes beyond those for which they were originally created. This worldview has implications beyond the forging of new code and technical artifacts.

Hacking and FLOSS participation often take on meaning as communal and shared actions.[13] As Jannie, a volunteer in the Netherlands, put it, "A big part of these groups is social. In that way we are like church groups."[14]

10. This heading title is an allusion to Latour's (by way of Archimedes) "give me a laboratory and I will raise the world," the idea that Pasteur built the strength of the laboratory by inducing and enrolling other elements of the world (like microbes and inscribed information) to run through it (1983). But I am not making a Latourian argument; hacking implies emergence (Kelty 2008).

11. Kelty 2008: 2.

12. Banks 2015; Coleman 2015. To *own*, *Own*, or *pwn* a server is to gain the top level of access and have free rein to do what you like (Coleman 2015: 160).

13. Coleman 2012a; Dunbar-Hester 2014.

14. Interview, Jannie, April 27, 2010, Amsterdam. Though many hackers are secular and atheist, their reverence for their activities often approaches religious fervor. Sociologist of hacking Sarah Davies writes that in her early research, she "toyed with the idea, eventually rejected, of using

Anthropologist Gabriella Coleman has demonstrated that hackers deploy a range of political stances including agnosticism and denial of formal politics (exceeding software freedom),[15] though implications for intellectual property in particular are at least implicit and often explicit in the technical and social practices of hacking.[16] Scholars have noted that the denial of formal politics makes FLOSS an unlikely site for gender and diversity activism, at least historically.[17] But FLOSS projects are not monolithic, and have matured over time. Arguably, the diversity advocacy that is the subject of this book represents a turning point within the collectivities whose focus is on FLOSS. As will be elaborated in the following pages, the shared enthusiasm for hacking and crafting code that unites FLOSS communities has collided with a realization that to believe that these communities are open in an uncomplicated way is naive. The communities have initiated debate and hacks of their dynamics, and there is no turning back, but these matters are far from settled internally. FLOSS projects and hackerspaces are also in dialogue with the wider culture, which is awash in "women in tech" discourse (including the high profile of Facebook chief operating officer Sheryl Sandberg's 2013 book *Lean In*). The raft of open-technology initiatives around diversity must be placed within this context, while keeping in mind that geek politics exist along a continuum.

Computing has become associated with freedoms, particularly notions of autonomy, self-actualization, and higher selfhood.[18] Common notions of what is at stake in open technology can be seen in this statement: "'Free software' means software that respects users' freedom and community. Roughly, it means that *the users have the freedom to run, copy, distribute, study, change and improve the software.*"[19] In other words, free software is here imagined to support users' autonomy, including expressive individualism. Users' autonomy is sometimes, but not always or even mainly experienced or expressed at the individual level; the community, group, or

---

the literature of the sociology of religion—conversion narratives, evangelism, the construction of higher purpose and meaning—to analyse involvement in hacker and makerspaces" (2017c: 225). Thanks to Christo Sims for discussion on this point.

15. The Free Software Foundation explains, "To use free software is to make a political and ethical choice asserting the right to learn, and share what we learn with others. Free software has become the foundation of a learning society where we share our knowledge in a way that others can build upon and enjoy" (Free Software Foundation n.d.).

16. Coleman 2012a. See also Kelty 2008.

17. See Nafus 2012; Reagle 2013.

18. Kelty 2014: 214.

19. GNU Operating System, n.d. Emphasis in original.

project is also a meaningful unit, and its freedom is also important. In North America, this might be the freedom to tinker with a given program in order to modify it to suit one's own desires or to scratch an itch.[20] In Latin America, this might be the freedom to not be subject to intellectual and economic constraints imposed on one's government or education system by a foreign corporation.[21] These freedoms are related and exist along a continuum, but they are not identical interpretations of the scope and mandate of open technology.[22] The articulations of freedom within the FLOSS community matter because at present community members are challenging their cultures from within: How can commitments to freedoms be reconciled with the unequal treatment experienced by some members of the community? For all the rhetorical attention paid to individual freedoms, hacking is suffused with collectivity; diversity and inclusion work insists on *bringing collectivity to the fore*, making self-consciously collective worlds.[23] It is significant that the hacking- and FLOSS-inflected rhetoric of freedom is always ready to hand for those attempting to hack their communities; as one person wrote on a feminist hacking list, "[our] actual goal is freedom from the undesirable attitudes which in partriarchy [*sic*] are hitched to gender."[24]

Hacking Emancipation

In 2006, prominent legal studies scholar Yochai Benkler buoyantly claimed that the "networked information environment" that materially and ideologically undergirds FLOSS "enhances individual autonomy" by "improving [individuals'] capacity to do more for and by themselves; [it also] enhances their capacity to do more in loose commonality with others, without being constrained to organize their relationship through . . . traditional hierarchical models of social and economic organization."[25] He went on, "Individuals are using their newly expanded practical freedom to act and cooperate with others

20. Raymond 1998.

21. See Chan 2013; Takhteyev 2012.

22. As a matter of historical and genealogical accuracy, it is worth noting that there are semantic and ideological distinctions between free software and open source. Yet as they are mobilized by open-technology communities seeking to consider diversity issues, they are more alike than different, which is why I lump them together in this book without hesitation. The term *open technology* encompasses both.

23. Thanks to Sarah Myers West and Mike Palm for comments here.

24. Email,—to [Feminist Hacking List], November 1, 2018. Thanks to Mike Palm for discussion.

25. Benkler 2006: 8.

in ways that improve the practiced experience of democracy, justice and development, a critical culture, and community."[26] Benkler exults in FLOSS participants' "newly expanded practical freedom" and the lack of "constraint" experienced by individuals in FLOSS cultures, and flat out rejects the notion that traditional lines of social organization could be relevant in FLOSS cultures. For him, the improved experience of justice and democracy is without question open to all; FLOSS not only levels social hierarchy but consigns it to irrelevance in all networked modes of collaboration. Coleman uses more tempered language but makes a congruent claim: "The arena of FLOSS establishes all the necessary conditions (code, legal protection, technical tools, and peers) to cultivate the technical self and direct one's abilities toward the utilitarian improvement of technology. . . . FLOSS allows for technical sovereignty."[27] This is a worthwhile summary of the scope and aims of FLOSS, and even open technology more generally, where an emphasis on code may be less central (though open-source software often supports open hardware).[28] Coleman's account is one of the richest cultural studies of the FLOSS lifeworld, enlarging our understanding by accentuating hacking in its idealized form. The backdrop of the ideals that animate FLOSS, against which its shortcomings stand out, sets up the contestations that form the subject of this book.

It is probably no coincidence that as celebration of open technology reached a fever pitch in accounts like Benkler's, diversity advocacy in open-technology cultures also began to heat up. At the heart of the contestations around diversity in hacking and opening up technological participation lies the fact that a substantial appeal of hacking has to do with *agency*. "The emancipatory potential of hacking exists precisely in that it crosses the line of who can access technology," writes Johan Söderberg.[29] The topic of this book is the relatively new but broadly accepted notion that the promotion of diversity in tech[30] is a social good, worthy of attention and advocacy, as well as the exploration of the myriad conceptions of what is at stake for its champions in open-technology cultures. Once again, we should note a

26. Benkler 2006: 9.

27. Coleman 2012a: 120.

28. Powell 2012.

29. Söderberg 2008: 30.

30. Here I use *tech* as a shorthand for development of software, electronics, and computing hardware, following media, education, and industry usage. I do argue however that an all-encompassing notion of *tech*—particularly analytical looseness about tech *participation* and *participants*—serves to reproduce rather than destabilize existing hierarchies and power arrangements.

disconnect between agency as an abstract individual capacity and diversity advocacy, which is necessarily socially embedded.

Access and emancipation are politically charged ideas: they offer liberal subjects inviting opportunities for self-determination as individuals and as collectives.[31] Technology, and computing technology in particular, has a tangled relationship to these political constructs, but in our present age, talking about the machines is never just talking about artifacts decoupled from these political valences. What is worth sustained analysis here is the ways in which technologists identify the tools and techniques that they find particularly emancipating themselves—here, computing, electronics, and related skills. Notably, they continually assert the conjoinment of computing, freedom, and progress. In the words of one prominent diversity advocate, whose technical background included a high-status programming role in a free software operating system project, "Open source was attractive because I never wanted my work to be thrown away. My job [ideally] has to have some sort of transcendent purpose. I loved puzzle-solving, making things work. If there's a problem with a computer, it's because you told it to do something wrong. [Working with code] can bring pleasure, adrenaline, and joy."[32] She neatly articulates many of the core beliefs of open technologists and computing devotees in general. First, the idea that FLOSS has a "transcendent purpose" (in contrast to contract work for industry or government entities); it lives on in the user base, whose members breathe life into it through use and modification. Second, the idea that the computer is a site where desire may be consummated in a joyous, almost formally aesthetic way, though it is also pleasurable in an embodied way to solve puzzles and "make things work."[33] Third—and this is a crucial point to note—her belief that if the computer doesn't work, it's because the user told it to do something wrong. Her quote exemplifies key elements of the belief system that for hackers[34] fuses computing to joy and emancipation.

This belief system animates hackers' extreme fervor for and commitment to computing technologies. It is not a stretch, therefore, to comprehend why

31. Kelty 2014.

32. Interview, Liane, July 24, 2014, San Francisco, CA.

33. See Streeter 2010; Coleman 2012a; Kelty 2008.

34. Here I am using *hacker* in an expansive sense to indicate users who have much closer relationships with computers than those held by average users, which are usually constituted in part by affective connections to this technology. I am using it in spite of the limitations of hacker identity, discussed in the next chapter, and sidestepping the common sensationalist media representation of the hacker.

those who experience these relationships with technologies would identify expanding the ranks of this form of participation as a crucial means of expanding agency for others as well. Hackers (including diversity advocates) are commonly extrapolating from their own experiences to articulate the notion that freedom and progress *for others* must also be related to access and emancipation through computing. As Christopher Kelty writes, "these tools engage our individual capacities to think, create, and manipulate the world, and they transform the collective relationships we have with others."[35] Participation in technical cultures and open technology in particular has been routinely hailed as attractive, even transformative. Indeed, one of the reasons open technology has been so celebrated is because (in conceptions like Benkler's, and countless others) it is held to offer its participants an opportunity to rework social relations, implicitly or explicitly contributing to their empowerment. Fundamentally, expanding this participation is often touted as a shortcut to enhancing political agency for everyone, and in recent years *particularly* for those groups who have been relatively sidelined vis-à-vis technological and political agency. Pursuit of technological emancipation is celebrated as an end run around discrimination and other social constraints.

Though these belief systems are persuasive for those who subscribe to them, we might ask what the consequences are when these beliefs are projected elsewhere. In particular, what are the entailments of exporting a progressive belief in the individual and collective power of technical agency to sites of social and economic inequality? Before we begin to answer this question, however, we should spend a moment unpacking how *technological* progress and *human or social* progress came to be imbricated.

Technology and Social Order (Or, How Hacking and Emancipation Came to Be Linked)

Historian of technology Leo Marx has persuasively argued that technological development was not initially bound to human progress; while it could be in the service of human progress, it was not interchangeable, at least in the early American republic. Importantly, the term *technology* was not widespread, either—its emergence as a prominent term with its present-day import did not occur until approximately the turn of the twentieth century. Preceding terms (such as *machinery*, *the mechanic arts*, and even the word *technology* itself, which did exist but narrowly referred to branches of learning

35. Kelty 2014: 198.

surrounding the mechanical arts) did not carry the same moral charge that today's *technology* does.[36] Indeed, according to Marx, *technology* acquired its potency in order to fill a semantic void, a nineteenth- and twentieth-century cultural longing to describe a novel form of human power that "the mechanic (or useful) arts" were insufficient to encompass, given their association with artisanal labor and individual-scale handiwork. It is this inflated notion of technology we are encountering when the large-scale technological society is imagined (no matter whether this is invoked as a freeing or frightening specter).

It is important to take technology seriously as an object of analysis. This is not nearly so straightforward as it may seem, as technology is as much an ideologically charged domain as it is a mundane artifactual component of everyday life. According to Marx, technology is a "hazardous concept": in our present society, it cannot help but *to stand in for things greater than artifacts*, and it is understood to have profound effects on social order. Conversations about technology are rarely about *artifacts in themselves* (though this phrasing may mislead us into thinking that a clear demarcation of technology from society, from power, and from social order is even possible, or desirable). Many critics of technology and culture have observed that stories told about technology reveal as much about the tellers as about the artifacts, and this is no less true here.[37] For these reasons, technology is a special case for social analysis: it is no less a product of social relations than other domains of culture, but its stature is so great and its shadow so long that it is worth concerted attention.

Marx's account offers a useful and sophisticated contextualization for hackers' and other enthusiasts' technological zeal as well as for the cultural baggage that accompanies this zeal. Hackers pursue technological development in order to maximize "human flourishing through creative and self-actualizing production."[38] Likewise, technology is a unique domain for the discharge of political energies. In the collective imagination, it has been vested with the power to initiate change (even as this belief obscures the role of social and economic relations).[39] Many technologists, especially those in activist geek circles, are motivated by political concerns and seek to build technologies that they believe can shift social power and redress social

36. Marx 2010: 562–64.

37. Sturken and Thomas 2004.

38. Barton Beebe quoted in Coleman 2012a: 15.

39. Marx 2010: 577.

imbalances or inequities. This certainly is true of quite a number of the people in open-technology communities whose efforts are under consideration here. Others, however, believe that "progress depends on the constant expression and reworking of already-existing technology."[40] In other words, the belief that technology is a progressive force outside of political channels is widespread among open-technology enthusiasts, though emphasis can be placed on different aspects of technology development, in particular developing (or reworking) technology for specific political reasons versus keeping technology under development for its own sake.[41] Lastly, and crucially, technologists often feel that dealing with technology offers a more concrete site in which to negotiate power and privilege—that is, it can seem cleaner to work on "technological solutions" than to wade into wider social contestations.

These points all explain why diversity advocacy has taken hold in open-technology cultures. Some community members emphasize the pursuit of both explicit and inchoate political outcomes through technological development. Others view the emancipatory experiences they have experienced working and playing with computers as worthy of export, and a singular and meaningful way to enfranchise people who have been relatively more excluded from various forms of civic, political, and economic citizenship. Often, these views overlap and serve to multiply adherents' commitments to diversity.

Pushback against diversity advocacy does exist, though it gets relatively little attention in these pages. Some can probably be attributed to naivete, the belief that these cultures and practices are already fully open to those who wish to participate. And it is undoubtedly the case that some opponents of diversity advocacy believe in "agency for me but not for thee," for reasons having to do with consolidating their own social and technical power, more or less consciously. Lastly, some articulations of freedom and meritocracy are incompatible with diversity work—for some, this work is seen as inherently nonmeritocratic and thus at odds with the cultural values of the FLOSS community.[42] In any event, the focus in this book is the internal

40. Coleman 2012a: 119.

41. Neither of these beliefs is inherently democratic, and both can be quite technocratic.

42. Thanks to Sarah Myers West for help with this point. Notably, the coiner of the term *meritocracy* intended it as a satirical concept, which was lost on many who uncritically adopted it in subsequent decades (Young 2001; thanks to Peter Sachs Collopy for this reference). At one feminist hackerspace, the wifi password when I visited was "meritocracy is a joke," pushing back on the meritocratic ideal, but only within the subaltern counterpublic of members and vetted

negotiations of diversity proponents and their mediating work within their open-technology communities.

Why does it matter if hackers and open-technology enthusiasts promote expanding the ranks of hacking as a means to wider political emancipation? What is at stake in framing proficiency in computing as a significant path to social inclusion? What is obscured in framing technical cultures as the appropriate site of intervention? And what does calling this *diversity work* accomplish (or fail to accomplish)? This book does not argue for diversity in technical cultures directly. Rather, it is interested in carefully investigating the political implications of diversity advocacy. Even in spite of advocates' often clearly emancipatory objectives, some of their framings of problems and solutions contain the potential to crystallize patterns of power that contravene their intentions. Across the chapters of this book, I argue that diversity advocacy has the most potential to change the expressive culture of open-technology communities. Interventions in these sites can serve as stages in miniature where people confront wider social problems; it can be powerful and galvanizing to strive for inclusivity in a social milieu over which one has some control.

At the same time, change in these communities is challenging, and not necessarily a stand-in for the broader change they hope to see. Though hacking is an exercise in world making, hacking has historically been a significantly different project from the one that diversity advocates are now challenging their communities to undertake. There is a potential disconnect between these local interventions in voluntaristic spaces and the wider, loftier effects that are supposed or hoped to flow from them. In other words, it is not enough to act locally while thinking globally—there are structural forces at work that dictate that these hacks will fall short of advocates' most elevated intentions. This is not to suggest that these interventions are worthless, just that their proponents are up against entrenched, monumental patterns. Partly, this confusion stems from the ways in which technology is hazardous: artifacts and artifactual production can wind up standing in for, or being confused with, social order. But voluntaristic communities, by their nature, are bounded. Open-technology communities are formed around a

visitors (Fieldnotes, July 2014, San Francisco, CA; see also Skud 2009; Reagle 2017). Amy Slaton has illustrated that meritocratic framings in engineering education were used to deny the fact that race could function as a determinant of students' life experiences (2010: 171). Finally, meritocracy can also be mobilized to argue *for* diversity initiatives (see Fowler 2015), though this is much rarer than "colorblind" (and the like) meritocratic framings.

shared enthusiasm for hacking. Challenges of social structure or systemic inequity are a heavy lift for DIY communities, and communities constituted around technology face unique challenges.

Without referencing these social issues directly, Mel Chua, a self-described "contagiously enthusiastic" hacker, writer, and educator touches on some of them. The social relations and historical patterns that surround computing and hacking are always freighted, and often reflect the priorities and interests of groups with greater social power, including elites, technocrats, and corporations. Chua writes in a 2015 blog post,

> As a kid, I *wanted* to choose the privilege of being oblivious and keeping my head down and immersing myself into the beauty—the sheer beauty!—and joy of STEM for STEM's sake . . .
>
> But I *couldn't* "just geek out about nerdy stuff." The environments where I was trying to "learn about nerdy stuff" were sociotechnically broken in a way that made it hard for me (as a disabled minority woman, among other things) to join in. If I wanted to even *start* being part of the technical community, I had to start by *fixing* the technical community—patching the roof and fixing the plumbing, so to speak—before I could even walk inside and start to live there . . .
>
> It was as if I could only enter the makerspace as a janitor.[43]

To paraphrase Chua, the aesthetic beauty and agentic participation hailed as attractive attributes of technical engagement were less available to her. Even though she showed up excited to partake, there were barriers to her "walking into" the open-technology space, let alone to her "living there." She experienced, firsthand, some of the historical patterns that historian of technology Amy Slaton has called the *relational* nature of science and engineering knowledge, and of technological skill and talent: "All of these relations involve the constant making of knowable students and employees by those with influence."[44] In other words, science and technology have historically been sites for cultural sorting work, separating STEM-capable people from STEM-incapable people. Slaton's analysis suggests that moving some people from one category to another does not destabilize the use of STEM as a site for this kind of problematic cultural sorting.[45] Chua's analysis is more ambiguous: What effect does "patching the roof" have

43. Chua 2015. Emphasis in original.
44. Slaton 2017.
45. Slaton 2017.

beyond a single building? To extend her metaphor, is the plumbing connected to sewer systems and water treatment facilities that taint the tap water?

Scale is an issue here: diversity advocates hope to rectify social problems that are deeply entrenched, which span sites such as higher education and industry, not only their own more intimate voluntaristic spaces. But their own social worlds, closer to home, are composed of people who are led to join by their shared enthusiasm for technology. Their preferred solutions are to hack their projects and cultures—to patch the roof and fix the plumbing. This comes up short as a solution for the problem of unequally distributed social power. In other words, distributing diversity in technical participation is not equivalent to generating justice—and it can never be equivalent. In fact, cultivating diversity without a robust critique of power can wind up placing open technologists' efforts adjacent to the goals of industry and neoliberal government initiatives.

But another problem is conceptual or definitional, having to do with how social problems are framed. If the goal is to distribute diversity, to sow more different kinds of people in technical production, then diversity advocacy is plainly making some inroads, though advocates and other community members may quibble about whether group X or group Y is proportionately represented in project A or project B, etc. But as feminist theorist Sara Ahmed argues, "diversity" inheres in individuals: "diversity is what individuals have *as* individuals."[46] Moving the needle on the individuals' diversity in a given project or institutional setting "gives permission" for people to turn away from institutional or societal inequality.[47] Put differently, to frame social inequality as a question of diversity in technological production, and to expect to change wider inequities by adding "diverse" individuals to technical cultures, is to misunderstand how the distribution of various social identities in a given sector are *outgrowths of differential social power*, not the other way around.[48] As political theorist Joan Tronto writes, "the process by which we make some questions central and others

46. Ahmed 2012: 71. Emphasis in original.

47. Ahmed 2012: 71.

48. Harding 1995. In her foundational study on early communities in cyberspace, Lori Kendall writes, "The culture of [her research site] and similar online spaces have been constructed by people from particular (relatively homogeneous) backgrounds. As such, these cultural contexts continue to appeal to people from those backgrounds and to re-create particular meanings and understandings. *Increases in online diversity will not necessarily change these existing norms*" (2002: 216, emphasis added).

peripheral or marginal is not simply a benign process of thought."[49] In social analysis and intervention, where the borders of care are drawn is critical.

The project of this book is to name and elucidate these dynamics in order to help advance a political project of justice in a technology-oriented world. It is important to recognize and validate the critiques of power that are emanating from open-technology communities. But what this book suggests is, if technical enthusiasts are experiencing their consciences being stoked vis-à-vis unequal technical participation, they stand to gain from placing their critiques into productive dialogue with conceptions of these problems that can help illuminate the ways in which STEM participation both is and is not contiguous with social and economic power writ large. To have a clearer understanding of what various concerns and interventions around diversity can and cannot accomplish is to set the stage for confrontations that may allow more concerted, less prefigurative changes to become possible.

If Diversity Is the Answer, What Is the Question?

Up front, it should be noted that I do not attempt to define diversity or hold fast to any particular definition in this book. I treat the concept as an "emic" one, emanating from within the communities that form the subject of this study. The questions I ask are: Why is diversity so important? and, What work is diversity doing (or meant to do) in these cultural spaces? As Sara Ahmed writes, the "mobility of the word 'diversity' means that it is unclear what 'diversity' is doing, even when it is understood as a figure of speech."[50] In other words, even though we understand what diversity means, it is not clear what work it is doing, or is meant to do. Her observation rings true here, and thus the book's agenda is to map and analyze where diversity has traction, and what work it is doing. The mobility and the limits of diversity are of primary significance in this book. Sometimes, its ambiguous meaning is part of what makes diversity work as well as it does; it is shifting and nebulous, ripe for appropriation in different contexts, with a protean (and contested) political valence.[51]

The mobility of *diversity* is easy to grasp, as it is a ubiquitous term and concept. Our contemporary moment is saturated with exhortations for women and members of other underrepresented groups (but particularly

49. Tronto 1993: 4.

50. Ahmed 2012: 58.

51. Thanks to an anonymous reviewer for helping draw this out.

women) to take up participation in STEM. "*Building up young people of color in tech* so that we can finally tackle structural inequity and disparity in this critical industry . . . [is] integral to the advancement of social justice in America,"[52] writes Dream Corps, commentator Van Jones's "social justice accelerator." At the time of this book going to press, the National Center for Women & Information Technology (NCWIT) reports that 26 percent of the US computing workforce are women and less than 10 percent are women of color; 5 percent are Asian, 3 percent are African American, and 1 percent are Hispanic.[53] Rationales for this push to increase STEM participation vary, but common ones are national competitiveness and women's economic empowerment. (NCWIT also claims that by 2026, 3.5 million computing-related job openings are expected, and that at the current rate only 17 percent of these jobs could be filled by US computing bachelor's degree recipients.) Both of these rationales could be found on the Obama White House's website in 2015: "Supporting women STEM students and researchers is . . . an essential part of America's strategy to out-innovate, out-educate, and out-build the rest of the world"; and "Women in STEM jobs earn 33 percent more than those in non-STEM occupations and experience a smaller wage gap relative to men."[54] The Trump administration removed this page but also touted a memorandum to increase STEM education funding.[55] In Canada, Prime Minister Justin Trudeau claims to be building a feminist government, mandating diversity and inclusion frameworks under all government policies and programs.[56]

Industry, too, often regards increased women's participation as desirable. Google neatly summarizes a 2015 corporate agenda surrounding women in technology fields on a webpage: "Technology is changing the world. Women and girls are changing technology . . . We always believed that hiring women better served our users."[57] In other words, the corporation's full

52. Email, Dream Corps mailing list, February 16, 2019. Emphasis in original. Thanks to Chenjerai Kumanyika for discussion.

53. National Center for Women & Technology. I use *Hispanic* here because NCWIT does. See Margolis 2010.

54. "Women in STEM," n.d. The page also quotes President Barack Obama as having said in February 2013, "One of the things that I really strongly believe in is that we need to have more girls interested in math, science, and engineering. We've got half the population that is way underrepresented in those fields and that means that we've got a whole bunch of talent . . . not being encouraged the way they need to."

55. United States, 2017.

56. Prasad 2018. Thanks to an anonymous reviewer for pointing this out.

57. http://www.google.com/diversity/women/index.html, accessed February 2, 2015. Another Google page additionally stated, "Our goal is to build tools that help people change the

market potential is not being realized without a developer base that can cater to diverse users. On another page, entitled "Empowering Entrepreneurs: Our Future," Google explicates the global reach of its vision and reiterates that technology is a route to empowerment: "Archana, an entrepreneur from Bangalore, shows how women are using technology to better their businesses, improve their lives and make their voices heard around the world."[58]

In many ways, the diversity advocacy that I examine in this book bears similarities to government and industry agendas. But unlike White House policy or Google programs, the initiatives I examine are driven by the voluntaristic ethos that surrounds FLOSS. We have to explain why fairly grassroots civil society groups also are pouring their energies into this diversity advocacy, often as volunteers. Diversity advocacy here is not necessarily identical to corporate, higher education, or government agendas, though there is certainly overlap. To tease out these similarities and differences requires careful parsing of the values and import vested in open technology.

It is also important to note that the *who* of diversity is flickering, not holding fast to a single definition or category of people. We might ask, whose diversity is most symbolically or strategically important, and why? In the earliest and widest instances of diversity advocacy, the *who* of diversity usually means *women*, as above examples illustrate.

This is not, of course, the extent of diversity that mattered and matters to advocates. Revisiting the comments by Mel Chua, above, we see a departure from gender as a primary identity category, as she invokes not only being a woman but also her status as a disabled person and her membership in a minoritized ethnic category (she is deaf and self-identifies as Chinese–Filipino American).[59] This represents a deepening understanding of "diversity," including a perceived need to be more intersectional.[60] By that same token, it represents the fact that many advocates found that agitating for "more women" was inadequate as a diversity goal. In other words, if diversity meant more women, for example, what did that leave out? Perhaps people from racial and ethnic backgrounds less likely to be present in

---

world, and we're more likely to succeed if Googlers reflect the diversity of our users" (http://www.google.com/diversity/women/our-work/index.html, accessed February 2, 2015).

58. http://www.google.com/diversity/women/our-future/index.html, accessed February 2, 2015. Note that while my research sites are mainly in North America, Archana is in India; technical work is used to bring people into globalized capitalism, literally and figuratively (Freeman 2000; see also Qiu 2016).

59. Chua n.d.

60. Combahee River Collective 1977; Crenshaw 1991.

open-technology communities? Or an understanding of gender that is less binary? People outside of North America and Europe? Advocates are not wrong to draw attention to the lack of representation of various groups in open-technology communities.

Going further, though, how do these multiple framings of identity within open technology serve to produce a politics of representation? What are the consequences of this act of production? What does a politics of representation fail to capture? This book argues that the current advocacy around diversity in open technology, with its emphasis on identity categories, largely circumscribes core questions about social and economic power that are suggested by advocates' engagement with diversity in open technology. An issue in need of recognition is that many advocating for diversity are located in North America and Europe—that is, in the Global North. In advocating for more women, a more expansive notion of gender, or more members of minoritized racial and ethnic groups in open-technology communities, advocates overlook the fact that there is a global underclass whose work materially supports the productive power of open technology.[61] In other words, the material and discursive output of FLOSS is quite literally made possible by labor that extracts raw material and manufactures hardware, which allows FLOSS and hacker communities' technical engagement. If we zoom out from the Global North and take an expansive notion of tech work—including the labor that undergirds hacking and open technology—it can hardly be said to have a diversity problem per se, because women workers of color actually abound. Thus, the diversity problem with which advocates mainly struggle must be seen in context as an attempt to *expand the ranks of an elite position within global capitalism*—high-status, well-paid tech workers. The diversity advocacy that forms the subject of this book attempts to change the constitution of open-technology communities while struggling with its ability to realign the social and economic power relations in which open-technology work is implicated.

As stated above, I do not attempt to define diversity or claim that one single definition of this concept suits the work I am doing here. Instead I regard it as a keyword that emanates from open-technology communities (though it does not originate with them, of course). I am interested in the work that it does in the social imagination of amateur technology cultures

61. Qiu argues that, far from being new, shiny, and digital, these conditions are an outgrowth of old industrial geopolitics (2016).

centered on "open stuff."[62] In my conception, it is precisely the murky outline of diversity that allows it to attain power, especially as it is not a value to which many people are easily opposed.[63] In some ways, diversity advocacy in these sites simply mirrors wider roil about less-than-equal standing and mistreatment of minoritized people in a variety of settings, which has become visible in campaigns from the life-and-death stakes of the Black Lives Matter movement[64] to feuds about representation embodied by, for example, the #OscarsSoWhite campaign about race and inclusion in Hollywood. But this centering of effort around a *technical* domain is significant and singular because of how technology is understood as a special and potent site within our culture.

It is essential to keep in mind the history of science and technology being touted as universally accessible and meritocratic *while in practice serving as sites of social sorting*, as indicated by Slaton. As Ahmed argues, "adding color to the white face of the organization *confirms the whiteness of that face*"[65]; decentering the dominance of certain groups requires more than simply adding members of minoritized groups. This seeming paradox may partially account for the ambivalence that some people of color feel with regard to the overtures of would-be white allies around diversity in tech (discussed in chapter 7). It also underscores that while attention to diversity is necessary, it is far from sufficient for an antiracist social justice agenda. In her research on low-income women's experiences with the "high-tech future" promised by digital technology, Virginia Eubanks describes a woman with a computer science degree who worked as a bus driver, because her son was disabled. Her responsibilities for care meant that she was unable to retain a position in the high-tech field for which she had been trained.[66] This kind of care work falls disproportionately to women and mothers. In drawing together these observations, I do not mean to suggest that diversity advocacy in open-technology communities is hopeless or without merit—only to illustrate

62. Skud 2011.

63. Exceptions exist: when diversity is interpreted as affirmative action or as at odds with meritocracy, opposition in FLOSS can be intense.

64. Rickford 2016; Mislán and Dache-Gerbino 2018.

65. Ahmed 2012: 151, emphasis in original. Ahmed writes that "diversity pride" may demand solidarity with whiteness. On the other hand, Ralina Joseph explores the "strategic ambiguity" of "postracial" terms like "inclusivity" and "humanity," arguing that they can be a way of naming and resisting racism (2018).

66. Eubanks 2012: 75.

that the terrain upon which advocates stage their interventions is a top layer resting upon sedimented strata that, not unlike geological formations, have formed over time through immense force. Rather than assuming that diverse people cultivating diverse technologies will lead to a more egalitarian and empowering technological future, it is essential to keep at least one eye squarely trained on the social and economic conditions that group people and endow them with differential opportunities, and how *technology itself* is implicated in projects of social sorting and domination.

How This Book Is Organized

Across the chapters, the central themes in this book have to do with how diversity advocates define their borders of care (principally: they care about promoting justice and equality broadly construed, but tend to limit intervention to already-constituted and bounded technical cultures, with ambivalent results) and how they are continually confronted by problems of scale, in that they are seeking wider emancipation but are limited to hacking versus effecting deeper structural transformation. It proceeds as follows. Following this introductory chapter, chapter 2 provides a historical background of the cultural strands that intertwine to produce diversity advocacy in open technology. It gives an overview of the history of women in computing, cyberfeminism, and hacking and FLOSS, while challenging conventional accounts of hacking. Chapter 3 explores what diversity advocacy builds in terms of techniques of governance and sociality to support a subaltern counterpublic and to speak back to a wider collectivity of open technologists; it illustrates the painstaking local-ness of many infrastructural interventions. Chapter 4 continues to examine what is built on a more literal artifactual level, describing what is being produced in sites of diversity advocacy, including code and craft. It argues that the significance of much of this material production is symbolic identity work; care is manufactured as much as things. Chapter 5 examines diversity advocates' imaginaries of work and labor, many of which are contradictory, both aligning with and critiquing market values. This topic matters because, especially as advocates envision their practices as potentially promoting worker power, their analyses generally do not fully account for the protean boundaries of so-called tech work and actual, material labor conditions, including the lower-status labor that supports Global North hacking. Chapter 6 follows diversity advocacy as it intersects with political stances that relate to but are broader than diversity advocacy: social justice activism, antimilitarism, and anticolonialism. These are sites where feminist hackers in particular often articulate connections to broader values that can inform hacking (and vice versa), but often stop short of full-throated critique, in part because of the ambiguous relationship of their activities to paid labor (some of which is laid out in chapter 5). Chapter 7 explores social identity and multiple conceptions of who might embody the *missing diversity* in open-technology cultures, from the perspectives of diversity advocates. It discusses gender, race, and ethnicity; proposes that representation has its limits as a project of empowerment; and suggests, again, that workplace relations to some degree constrain the criticism that advocates express. Chapter 8, the conclusion, pulls together threads in the previous chapters to assess the potentials and limitations of diversity advocacy in open technology as a site for claiming equal rights, and as a quest for representation; it also evaluates the market logics that accompany this advocacy. Finally, it meditates on the challenges inherent in centering a project that insists on a redress of imbalances of power around technology, arguing for a project of justice and equity that ironically decenters technology as a primary axis of intervention. It argues that while voluntaristic tech communities cannot singlehandedly attain the scale of the endeavors they hope their interventions will address, they are well-positioned to offer care and analysis that can set a more expansive, yet more rigorous, agenda.

Research Methods

The focus in this book may appear somewhat difficult to define. Diversity advocacy in open technology is not a social movement in any traditional sense; neither is open technology itself. What this study attempts to track is the constitution of belief within a geographically dispersed and heterogeneous community, and in particular an impulse to intervene in that community. For many technologists, a "theory of change"[67] centers around technology itself: the belief that tools are drivers of social progress.[68] I situate diversity advocacy within what sociologist Anselm Strauss called a "social world perspective."[69] For Strauss, a shared social world is first and foremost discursive, bound together by communication, but also includes "palpable matters" like activities, memberships, sites, and organizations.[70] It is

67. Interview, Anika, July 7, 2015, New York, NY.

68. Thanks to Todd Wolfson for discussion on this point.

69. Strauss 1978.

70. Strauss 1978: 121. It would also be possible to think of diversity advocacy through the conceptual lenses of a social imaginary or publics (Habermas 1991; Kelty 2008; Taylor 2004). This

challenging for the analyst to bound social worlds, because they both intersect with other worlds and segment into smaller subworlds.[71] The construct of a shared social world maps quite well to the field of diversity advocacy, which overlaps and intersects with wider open-technology cultures (e.g., mainstream FLOSS) and even with the imaginary of the tech industry,[72] or Silicon Valley, but is distinct from each of them, or not wholly of either of them. In addition, diversity advocacy segments into smaller worlds: different political, technical, and affective strains coexist within it, sometimes working toward contradictory goals. Although they are more similar than dissimilar, it is important to remember that the sites and groups whose activities comprise diversity advocacy within open-technology cultures are not monolithic.

As Christopher Kelty writes, "What advocates, adherents and proponents of free software see in it is something more than software—they see a style of remaking the world, and they immediately want to apply it to every aspect of life."[73] As noted above, diversity advocacy derives from the same impulse to remake the world that FLOSS does. The practitioners in these pages hold a shared belief that there is a "bug" in how open-technology communities have been constituted, which will leave a corresponding negative imprint on their technical output. They thus believe that in fixing bugs in the community, they will attain better technological outputs, and thus a better social world; the technical and the social are imbricated and co-constitutive.

Diversity advocacy is quite evidently multisited and multivocal.[74] Methodologically, tracking and making sense of this social world is challenging, but I have attempted to conduct what might be called a situated genealogy of the present.[75] My research methods here are informed by an ethnographic

is most useful in thinking of its relationship with the wider public of FLOSS, so I lean more on this in chapters 3 and 4, especially Nancy Fraser's conception of subaltern counterpublics (1990). Thanks to Chris Kelty and Lucas Graves for sharpening my thinking on this.

71. Strauss 1978: 122–23.

72. It is erroneous to conceptualize the tech industry as a singular entity, as opposed to a series of related industries (including software, hardware, materials, streaming entertainment, e-commerce retail, and more) whose interrelationships are a moving target; see Jonathan Sterne's explication, "There Is No Music Industry" (2014), and also Stoller 2014.

73. Kelty 2013.

74. George Marcus discusses "multi-sited ethnography" as a way to adapt to more complex objects of study (1995).

75. Strauss called for linking fieldwork and interviewing to "historical and contemporary documentation" in social worlds research (1978: 127). Of course, referring to genealogy invokes Foucault 1995; see also Jordan 2016.

sensibility, but lack the "deep hanging out"[76] component that is a hallmark of traditional single-site ethnographies. Instead, I have sought to mirror the distributed nature of this advocacy, conducting participation observation at a number of sites (North American hackerspaces, digital fabrication labs, software conferences, "unconferences," corporate events, and software training events and meetups). An alternative approach would have been to embed myself and closely attend to a single FLOSS project or hackerspace, but the networked nature of this phenomenon means that to follow the actors I had to traverse multiple sites.[77] This might be called *polymorphous engagement*, in which I interact with actors across a number of dispersed sites, some in virtual form; this approach preserves the pragmatic amateurism that has characterized anthropological research (even as it moves away from a singular emphasis on participant-observation).[78] It also allows me to trace multiple simultaneous emphases and orientations within diversity advocacy; employing "comparative optics" [79] allows for a meaningful analysis of both common features that cohere diversity advocacy and different orientations within it.

Fieldwork and data gathering spanned 2011 to 2016, with continuous attention to electronic mailing lists and online traffic, and punctuated conference attendance and interviewing. This period is meaningful because it has seen several feminist hackerspaces appear as well as growing attention to diversity in mainstream open source. At the same time, it is a snapshot of an unfolding story with both a prehistory and a future that are outside the scope of the present research. It is significant that several initiatives that became research sites were born during this period; while this indicates that I have had my finger on the pulse of a meaningful social phenomenon, it also means that the objects of study were a moving target and hard to identify, which creates a methodological challenge.

I have interviewed participants in these activities as well as founders of hackerspaces, open-source software projects, and initiatives to promote women's participation in technology (around twenty-five semistructured and informal interviews), mainly in North America and a few in Europe. Pursuing "eclectic data collection from a disparate array of sources in many different ways,"[80] I followed much online activity, lurking on project lists

76. Geertz 1998.
77. Latour 1987.
78. Gusterson 1997: 116.
79. Knorr Cetina 1999: 4.
80. Gusterson 1997: 116.

and following social media, which again mirrors the fact that many of these efforts are coordinated and distributed across space, even when they also include local, static components such as hacker and maker spaces, or meetups based on a project or a programming language. Conferences, of course, are important for participants—and the researcher—for the ritual elements that occur when a community comes together, not only for the information that is transmitted within them.[81] Software and hacker conferences can also be occasions for confrontation, including controversy and the policing of behavior and boundaries within a community, which are of interest to those seeking to interpret the values and meaning-making systems of a community. In tracing these threads of activity, I gain the ability to map the meaningful (and contested) discourses that surround diversity advocacy, situating them within varying social contexts. It is not an exhaustive perspective on these endeavors, but it is not wholly idiosyncratic either; I trace multiple skeins of distinct and interwoven activity in order to draw out meaningful contrasts, and interpret the implications of these varying positions for the groups staking positions within the space of this advocacy.

Researcher's Position

My own subject position and social identity is implicated in this research. As a white, middle-class, highly educated, and literate person in North America I experienced relatively open access to and hospitable treatment in these communities and their conversations. Being a professor, my interest in these sites required little justification in most cases. That being said, my training, expertise, and commitments are those of the academy, specifically interpretive social science at the intersection of science and technology studies and communication research, not computer coding, geeking, crafting, hacking, NGOs or startups, or feminist activism. This meant I was an outsider in meaningful ways, and participants did not generally mistake me for one of their own.

Approaching middle age during this research, I was roughly a peer with or older than many of the practitioners I talked to, but not old enough to seem out of place (yet).[82] As a mostly able-bodied person, I was able to take

81. Coleman 2010.

82. Mothers and grandmothers are the ur-example for the technically unskilled; "could your mother understand this?" is an exceedingly common trope in evaluating whether a given topic or presentation is "too technical." For example, "Explaining Virtualization to Your Mom in 5 Easy Steps" (Fenton 2014).

for granted access to physical spaces. As a native speaker of English, which is a globally dominant language generally and for technical cultures, I was not excluded from or in need of accommodation in order to follow spoken or written conversations. Hailing from the United States gave me greatest access to groups and conversations located in North America. While the book follows some conversations involving Europe—and a very few traces from Asia and Latin America—it is largely bound in space to North America and the United States in particular, though it attempts to be critically reflexive about how it is situated.

Some of these sites are literally closed to people who do not identify as women, or who are men-identified. Most were explicitly genderqueer and trans-inclusive; some required that people identify as women "in ways that are significant to them."[83] This means that as a person who identifies as a cis woman, my gender is implicated in my ability to conduct this research; such strictures draw out quite plainly the fact that the conclusions I make here are situated. Whiteness marks me as a member of a dominant social group, which matters in both women-centered and mainstream spaces. My whiteness confers many advantages, but it also means that there were a number of spaces and conversations where I was not particularly welcome to visit with the intent of representing the people there in print. Many people of color have felt burned by white women and white feminists and need their own spaces for regrouping. I sought out some conversations about these topics, but especially as this research design lacked the single-site depth required to build comfort and trust, I consciously avoided pushing for my own inclusion. I sought to strike a balance that was not indifference yet offered considered respect for the dynamics of these attitudes.[84] This being said, gender is the primary axis of diversity advocacy, so the primacy of gender in this work is a faithful rendering of advocates' priorities, not a distortion.

Research Sites: Names and Anonymity

This is not an exhaustive account of diversity advocacy in open-technology communities, but it draws from a heterogeneous cross section of diversity

83. Geekfeminism.org 2014.

84. Keeping in mind how I and others are all situated within the matrix of domination and recognizing others' capacity to represent themselves and their experiences, this felt like an acceptable, if imperfect compromise (Collins 2000).

advocacy. This heterogeneity is what drove site selection. Sites include the Python programming language community, where advocacy around diversity was particularly salient during the period of this research;[85] the Ada Initiative (2011–15), which ran electronic mailing lists and a series of *unconferences* to support women in open technology; and a handful of feminist hackerspaces and hacking groups. In addition, various FLOSS projects, some named, make appearances. The community that built and maintains the Geek Feminism blog and wiki (circa 2008 to present) runs through many of these sites like a ribbon, so even though that community is not one I directly explore, its imprint appears on many of these sites. What the sites I visited have in common is that they are not especially institutionalized and are suffused by a voluntaristic ethos; I am interested in how self-organized initiatives address and challenge tech cultures with regard to the issue of diversity.

In research encounters, I disclosed my presence as a researcher. I sought written consent for formal interviews, per Institutional Review Board (IRB) guidance and best ethical practice.[86] I generally audio recorded interviews and took longhand notes, except in one instance where an interviewee did not wish to be recorded. An introduction when one joins an email list is standard, but because introductions are conducted once, it is possible that list members may have forgetten over time that a researcher was present, and it is also possible that newer list members were unaware that someone on the list was a researcher. Similarly, in group introductions in larger settings, it is possible that my presence as a researcher flew past quickly enough that people did not fully absorb it. I was more careful to alert people that I was conducting research in small group and especially one-on-one interactions.

Whether and how to identify research sites is a complex matter. In traditional ethnographies, the researcher might assign pseudonyms to the sites, and sometimes cloud locale as well (e.g. City Hospital in Large Midwestern City). Norms are changing such that it is more common to identify sites and

85. Callahan et al. observed this as well (2016: 575). Python is not itself a FLOSS project; it is a programming language that is popular among FLOSS projects and enthusiasts.

86. IRB approval was required before the National Science Foundation award (cited in the acknowledgments) could be disbursed, as is standard. I had three institutional homes during the period of researching and writing this book. I did not seek renewed IRB approval after the NSF grant administration had ceased. This was a bureaucratic lapse, not an ethical one, as I was confident that my research protocol had not changed, and that I would continue to maintain an iterative, rigorous, questioning stance toward the ethics of engaging with my research sites, which exceeded IRB strictures (and which, even so, did not cleanly resolve all my dilemmas). I also maintained active IRB certification required to supervise research.

informants by real names.[87] I have chosen to name sites named above for multiple reasons. This is ethically justified on the grounds that these sites are already positioned as public entities, conducting debate and advocacy which is by its nature public-facing as it shades into journalism and events in the public sphere. These communities of open-technology enthusiasts are not particularly vulnerable (though individuals within them may be) and are of relatively equal standing when compared to the researcher, in terms of social capital. They do not stand to lose by my disclosing that their activities form the basis for this analysis. Moreover, to change characteristics of these groups beyond recognition to those familiar with this terrain would likely impede my ability to represent them in their particularity and heterogeneity, which creates a dilemma.[88]

That said, the particular events I narrate and interpret are not usually occurring in full public view; thus I do not usually associate events with a named group in the text. Even when I do, or when the group or project can be inferred, I never relate individuals' actual names, or give identifying details about individuals beyond basic demographic information that is analytically relevant. This was negotiated with informants, many of whom were fairly willing to go on the record but were concerned about their words or opinions being misconstrued as representing the official policies or opinions of their employers or the groups in which they were active. While some consented to being named, giving everyone a pseudonym provides greater cover for those who were ambivalent or wished to remain more hidden. Interview subjects were offered the opportunity to generate their own pseudonyms, sparing me the dilemma of whether or how to disclose gender and ethnicity or nationality in pseudonyms. In not a few cases, people were concerned that identifying details like age bracket, ethnicity, gender, and geography would link them as individuals to their words, in which case we agreed upon broader categories to use that were still accurate but less identifying. Some with handles in hacking circles indicated that I should use these as their pseudonyms, but because these link them to their online identities as much as their given names would—sometimes even more—I elected to change everyone's names.[89]

87. McGranahan 2014.

88. See Stein 2010.

89. Kendall (2002: 242) offers a good discussion of this dilemma. She is correct that assigning pseudonyms to other pseudonyms does not provide perfect protection and is potentially controversial. In this research, I also ran into the problem of several interviewees who are active in diversity advocacy being quoted online in blog posts and articles about these topics. In those

I have also not identified hackerspaces, because they are smaller communities whose internal workings can be sensitive for members, and their real names are not analytically relevant.[90] Nonetheless, in all cases, this terrain is specific enough that a person knowledgeable of these sites might have a strong hunch. Plausible deniability is more realistic and more analytically faithful than absolute anonymity.[91] The greater priority for me was to provide a layer of artifice that protects individual people from having their true identities associated with the utterances and actions that I narrate. One consequence of these choices may be that the sites get flattened a bit; a different approach might bring to bear significant features in order to draw out contrast between sites. A single- or two-site study would provide more intimate portraiture. As I am interested in diversity advocacy at a more zoomed-out level, as a social-world phenomenon, this trade-off is acceptable.

With interview subjects, the researcher has a greater ability to control privacy. But the reality of online research is that archives and traces of online communication may persist online, and a diligent person who wished to look up conversations I quote and link them to real email addresses or social media profiles probably could accomplish that with some.[92] Another option would be to paraphrase all communication that leaves a trace, which I considered, but concluded undesirable, because as sociologist Arlene Stein notes, I might inadvertently blunt the power and essential features of the narrative.[93] I have thus erred on the side of allowing real people's voices to

---

cases, I maintain distance between quoted sources and the identities of people I interviewed and interacted with, which means that in a couple of cases individual people and projects are represented as multiples of themselves in these pages.

90. A few site-identifying details can be seen in images. Having secured permission to use the images and having considered whether anything overly sensitive or compromising is being revealed in my discussion of these sites, I concluded that using the images is ethically justified and more beneficial analytically than leaving them out. Some individual faces can be identified in photos, too; but this only associates them with diversity advocacy, not with their statements or other actions. Wherever possible I sought permission from people whose faces appear. I also sought permission to quote (without attribution) some email correspondence that was private when it was sent (forwarded to me by interlocutors). If I did not receive a reply, which was not uncommon, I made a judgment call.

91. Interviewees were offered an opportunity to review the passages in the manuscript containing their quotes, and one key informant read an early version of a chapter and offered further comment at that time.

92. As Anne Beaulieu and Adolfo Estalella write, "One of the fundamental implications of traceability [in online research] is that decisions about ethical practices such as anonymization may not be in the hands of ethnographers" (2011: 12).

93. Stein 2010: 4. See also Bruckman 2002.

stand in the text, out of respect for them and in the belief that their actual words can best represent their thoughts and opinions, which are analytically relevant. One of the sites had an explicit policy about quoting being permissible as long as individuals were not named (à la the Chatham House Rule). While this is not a blanket permission for all the sites I traversed, it speaks to the ethos of FLOSS, where both transparency and privacy are cherished values (with the right to choose whether or not to disclose being paramount). I have used it as a rule of thumb for sites that did not have published policies. In all cases, I have attempted to respect individuals' privacy while leaving analytically relevant features of groups and people intact enough for a coherent analysis and respecting stated preferences. These choices were made in good faith, weighing considerations toward both informants and the scholarly and ethical traditions that inform this work.

2

History, Heresy, Hacking

Hacking's origin is well understood, and its origin myths legion. Phone *phreaking*—using audio tones to make free phone calls, to the dismay of Ma Bell—is one activity that stands out in hacking lore. (The hacker publication *2600* is named for the hertz frequency used to control the phone system.) Another famed locus of proto-hacking is the Massachusetts Institute of Technology campus, where obsessed students stayed up nights building elaborate model train systems, stringing together novel electronics, and prowling the campus using its network of infrastructure tunnels.[1] Though these examples are somewhat divergent, they are more similar than they are different: both began in the late 1950s and placed adventure-seeking, curious, youngish white men at the center of the action. And, by extension, this is the subject position we imagine when we think of a hacker: located at the center of technical culture, the hacker stretches the limits of technologies to flex his curiosity and assert agency.

This genealogy is important as we examine how voluntaristic technical communities grapple with patterns of participation among their ranks. This book takes the perspective that technical communities and their activities are co-constitutive. In other words, changing who is there may also change practice, and vice versa. This disputes widely held assumptions that technical practices like coding or tinkering constitute a neutral palette.[2] It also argues

1. Coleman 2012a; Levy 1984.
2. See Slaton 2017.

that for those whose social positions deviate from the norms around which the categories of geek or hacker have been called into being, negotiation with these categories is required.

Geeks and Hackers

The *Oxford English Dictionary* (*OED*) defines *geek* as "depreciative. An overly diligent, unsociable student; any unsociable person obsessively devoted to a particular pursuit."[3] This usage goes back to at least the 1950s. The *OED* offers a more recent, 1980s, definition of *geek* as "a person who is extremely devoted to and knowledgeable about computers or related technology," and notes that "in this sense, esp. when as a self-designation, not necessarily depreciative." Another iteration of the term meant a carnival performer or circus freak. This usage is a bit earlier, with the *OED* listing 1919 for a carnival performer and 1935 for the colorful description "a degenerate who bites off the heads of chickens in a gory cannibal show." Precedent for both the circus geek and the academic geek is found in nineteenth-century usage of a foolish, offensive, or worthless person. *Geeking*, the verb form, also has a lineage that might surprise us. By 1990, to *geek out* meant to study hard, a denotation linking the phrase to *geek* meaning studious diligence. More specifically, the *New Hacker's Dictionary* links geeking out to technology, offering the following definition: "*Geek out*, to temporarily enter techno-nerd mode while in a non-hackish context."[4] But in its 1930s incarnation, the *OED* defines *geeking* and *geeking out* as "to give up, to back down; to lose one's nerve" (in addition to the meaning of performing as a circus freak). Geeking was originally equated with weakness and failure.

This inversion—from geeking as weakness to geeking as mastery—is worth scrutinizing.[5] Even as geeking signifies academic or technical potency, it retains hints of cultural ambivalence. Though geeks may now be celebrated as heroes, they are still characterized in popular culture and the popular press as physically weak, socially maladjusted, and outside of normalcy more often than not. Geeks may or may not differ from nerds. On this distinction, science and technology studies (STS) scholar Ron Eglash quotes novelist Douglas Coupland, who writes that "a geek is a nerd who knows that he

3. *Oxford English Dictionary*, "geek, n." March 2019. Oxford University Press. https://www-oed-com./view/Entry/77307?rskey=B1Eu4Q&result=1, accessed May 17, 2019.

4. Raymond 1996: 214.

5. See Dunbar-Hester 2016a.

is one."[6] In other words, self-awareness and embrace of one's geeky status are components of geekhood. Geekhood can be borne with pride, whereas nerds are just nerds, dweebs, losers.[7] In popular culture, *geek* can mark outsider status, or outsider status along with studiousness.

To explain this drift over time, we might look to the transition from physical labor to "knowledge work" that has occurred over the twentieth century. As cultural historian Anson Rabinbach explains, throughout the nineteenth century, society was understood to be powered and moved forward by bodies at work: "The human body and the industrial machine were both motors that converted energy into mechanical work."[8] By the late twentieth century, though, the *mind* was ascendant, at least metaphorically. Industrial and physical work still exist, of course, but they have been rendered invisible by a combination of offshore manufacturing in global supply chains and the discursive exaltation of managerial and intellectual work, which marginalizes service work and manual labor. Geeks have ridden this shift in the opposite direction, moving from a position of weakness and marginality to a position of greater relevance and influence. Human minds, not bodies, are understood to be the seat of power in late capitalism. That said, geeks have not moved into unambiguously hegemonic positions, and the geek body retains its status as a site of spectacle. While geeks no longer decapitate chickens with their teeth, they are often portrayed as gawky, puny, and bespectacled[9]—perhaps not monstrous, but still deformed.

The genealogy of *geek* is important for multiple reasons. Not only is it now centered on knowledge (especially arcane knowledge), it has transmuted from a term of insult into a more positive descriptor. Many people using *geek* to describe themselves and others use it in a fond, self-aware form of teasing and playfulness. As with other iterations of identity politics, geeks have laid claim to a title with a history as a term of disparagement in

6. Coupland 1996 quoted in Eglash 2002: 61, n1.

7. The *OED* notes that *nerd* has also acquired a definition as a person who pursues a "highly technical interest with obsessive or exclusive dedication." However, it is still more likely to be depreciative, and it is also more broadly defined as "an insignificant, foolish, or socially inept person; a person who is boringly conventional or studious." Much more could be said here. For example, the appearance of the "black nerd" in popular culture indicates that reclamation of *nerd* is possible as well. Significantly, this appropriation recodes *nerd* racially, tying African Americanness to intellectualism. It thus decouples blackness from primitivism, a linkage exemplified in musician Brian Eno's statement "Do you know what a nerd is? A nerd is a human being without enough Africa in him," quoted in Eglash 2002: 52.

8. Rabinbach 1992: 2.

9. See, for example, Jain 1999; Pullin 2011 for discussions of how prostheses mark disability.

order to gain power over its use, and they now derive strength from a label that had once been injurious to them. Geeks' embrace of this term now signifies their own uniqueness, their distinctness from the mainstream and commonality with each other.

Notably, *geek*'s acquisition of positive valence and in-group signification coincided with computing's rise in prominence over the past three or four decades. Computers have made a leap in the popular imagination from symbols of dehumanizing bureaucracy to intimate machines for self-expression and liberation.[10] Programmers, computing magnates, and hackers have catapulted into the limelight. This is evident in the stature and perceived social power of such figures as Bill Gates, Steve Jobs, and Mark Zuckerberg. Their technical "wizardry"[11] is an object of public reverence. At the same time, especially as *geeks* shade into *hackers*, they may be met with suspicion and ambivalence, as Edward Snowden or Julian Assange can attest.[12] This indicates that the freakish, threatening elements of geekhood may still be conjured. Programming exhibits a long history of conflict between practitioners and the institutions that employ them, including contestations over requirements for entry and over craft versus science status, as computing historian Nathan Ensmenger has shown.[13]

Wizardry is also gendered, of course. Technical masculinity precedes computing.[14] Historian of communication Susan Douglas locates amateur (ham) operators' work with radio as a site of reinforcement of ideas about masculine identity and technical competence in the early twentieth century.[15] She discusses how the tinkering work performed by men and boys, celebrated in the press, helped attenuate tensions between conflicting definitions of masculinity. Tinkering offered access to a masculine technical domain that was accessible and valued, and that stood in contrast to masculine ideals of ruggedness, strength, and plunder, which were becoming less accessible and less valuable. Douglas's account demonstrates that radio amateurs seized the new technology and interpreted it in a way that emphasized

10. Streeter 2010; Turner 2006.

11. Rosenzweig 1998.

12. Gregg and DiSalvo 2013.

13. Ensmenger 2010a.

14. Though early computer operators were women, historians of computing have shown that many of the skills and practices associated with running the first electronic computers were not associated with a particular gender. In any event, this early history did not unseat the association of technological skill with masculinity (Abbate 2012; Hicks 2017; Light 1999).

15. Douglas 1987, chapter 6.

masculinity and how the technology related to different gender roles; the technology was used to reinterpret masculinity itself. Electronics tinkering was a remarkably stable elite masculine hobby during the twentieth century, offering suburban men and boys both a masculine space within the domesticity of the home and training for white-collar technical professions.[16] But by the last decades of the twentieth century, the object of tinkering had begun to shift away from radio and toward computers. The continuity of tinkering as a masculine pursuit offers some clues about geek identity. Computer geeks (like the hams before them) are overwhelmingly likely to be white men (or youth[17]), often from middle-class or upper-middle-class backgrounds. Reasons for this likely include exposure to computing at a young age; parental educational achievement; gender expectations and socialization of children and youth; and cultural norms in computer science and hobbyist communities, among others.[18] Geek identity is one factor perpetuating exclusivity in some technical cultures ranging from engineering schools to Silicon Valley. Yet for all of its exclusivity, it is worth noting that geek identity represents a nonhegemonic form of masculinity:[19] geeks may feel persecuted (especially in high school) and geek spaces (online or off) have historically constituted sanctuaries for young men to revel in alternate forms of sociality, leaving oppressive elements of mainstream culture behind.

None of this is to suggest that participation in computing or related technical pursuits is closed to people who are not white men, of course. Strategies to combat the association of geekiness with white masculinity include linking geek identity to technical *engagement* as opposed to technical *virtuosity*.[20] Technical communities including free and open-source software groups and hackerspaces have repeatedly sought to address issues of diversity within their ranks, as this book demonstrates. Women can and do identify as geeks.[21] And Ron Eglash argues that Afrofuturism is an example of an improvised way to achieve technical prowess or identification without

16. Douglas 1987; Haring 2006.

17. See Coleman 2012a: 28–30 on youth and coding.

18. Kendall 2002; Margolis and Fisher 2003; Misa 2010. See Ensmenger 2010a on the historically tenuous status of programming and the rise of academic computer science.

19. Some have speculated that geek masculinity got diluted with the appearance of Wall Street and "brogrammer" types in Silicon Valley (Baker 2012), while not unseating masculinity in tech. I will leave it to others to sort out the veracity of these claims; see Rankin 2018, chapter 2 on fraternities and computing in the 1960s.

20. Dunbar-Hester 2008, 2010.

21. Newitz and Anders 2006.

being tied to geekiness per se.[22] Yet the association of white middle-class masculinity with aptitude and affection for computing is entrenched.

These dynamics may be easier to see in cases considered to be peripheral to hacking (or not considered to be hacking at all). A couple of instances are illustrative. Though leisure activities were a point of entry for whites into technical occupations, by contrast, historian of technology Rayvon Fouché has documented how the meaning of *tinkering* and *inventing* for African Americans was circumscribed by racism. Acts of black invention were framed in a negative context; whites declined to celebrate black ingenuity. Fouché quotes a nineteenth-century white lawyer weighing in on a dispute over a black inventor's patent application: "It is a well-known fact that the horse hay rake was invented by a *lazy negro* who had a big hay field to rake and didn't want to do it by hand."[23] In a similar vein, Mexican American lowrider car culture, while occasionally celebrated as an urban folk culture, has historically needed to defend itself against being associated with criminality, and has rarely been portrayed as hacking in a positive, agentic sense.[24]

Thus, returning to cultural ambivalence toward hackers or hams whose activities might at times be met with suspicion, it is apparent that whiteness has been a resource for white technologists who are understood as nonthreatening, "good" actors in society—more dazzling wizards than criminal threats. Electronics tinkerers who were white could cast even boyish mischief (as practiced by radio hams who trolled the US Navy with obscene messages in the 1910s[25]) as a stepping stone to cultural legitimacy in the form of high-status technical employment. Needless to say, such behavior would likely be met with harsher social or even legal sanctions for members of social groups whose expressions of dominance (technical or otherwise) would not be celebrated by the wider society.[26] Unlike their white counterparts, African Americans exhibiting technical ingenuity might not strive for subversion as much as for legitimacy; some were motivated by the goal of acquiring proper credentials to assimilate into white society.[27] Upwardly mobile middle-class

22. Eglash 2002. See also Everett and Wallace 2007; Mavhunga 2014.

23. Fouché 2003: 647 (emphasis Fouché's). See also de la Peña 2010; Fouché 2006; Green 1995.

24. See Chappell 2001; Henry 2013.

25. Douglas 1987, chapter 6.

26. In our present day, Arab American or Muslim hackers might be met with extreme suspicion. The 2015 case of Ahmed Mohamed, a Sudanese-American high school student in Texas who built a homemade clock and was accused by his teacher of making a bomb, brought forth charges of ethnic profiling and Islamophobia (Chappell 2015).

27. Fouché 2003: 6.

whites, on the other hand, did not need to assimilate, and their acts of pranking can be understood as an expression of not only playful engagement with technology but also social dominance.[28] It is not necessarily true, though, that either hams or hackers as a class would recognize this consciously.

When black hackers *do* occur in literature or popular imagination, as they did by the late twentieth century, the signs and signifiers of blackness are largely "used to empower subordinate identities (in this case, hackers) within white society,"[29] according to Kali Tal. In other words, while the blackness of these characters bolsters their connotation of subversion vis-à-vis mainstream white society, these representations do not necessarily traffic in *actual* black lived experience, but appropriate blackness in the service of bolstering the revolutionary or outlaw nature of hacking.[30] In a related vein, information studies scholar Lilly Nguyen posits that hacking in present-day Vietnam is best understood as breaking *into* global technology cultures from which most people in Vietnam have been excluded. This is in spite of the fact that hegemonic hacking is understood to be breaking *out of* social and technical limitations.[31] Nguyen points out that this definition of hacking only applies to hacking that is already integrated into established technical cultures. To understand the practices of Vietnamese hackers requires a shift in understanding of what hacking is.

All of this is to suggest that this book cannot invoke the history of hacking without troubling more traditional accounts that center around computing and electronics, which neglect innovators and appropriators of technology who fall outside the most common scripts.[32] This chapter offers an abbreviated history of hacking, open source, and open-technology participation that offers context for the activities described in this book. At the same

28. Control over technology has historically been used to enforce social order; technical expertise was constructed in the late nineteenth century in a manner that exerted the power of the "electrical priesthood" over women, rural and lower class people, African Americans, indigenous people, and colonial subjects (Marvin 1988).

29. Tal 2000.

30. And the presence of people like Art McGee, a Bay Area African American "cyber organizer" who maintained Afronet, an electronic bulletin board system (BBS) in the 1990s, is relatively overlooked (McIlwain 2019; see also Bailey 1996).

31. Nguyen 2016 (emphasis in original). See also Beltrán 2018. Gajjala (2014: 289) argues that marginalization itself is reproduced unequally; the "very point of access to the global" can become the point of marginalization for those from non-Western spaces.

32. See Eglash et al. 2004; Nakamura and Chow-White 2011; Tu and Nelson 2001. For historical examinations of race and racialization in computing, see also McIlwain 2019; McPherson 2011; Nakamura 2014; Nelsen 2017. Benjamin (2016) underscores that technoscientific innovation is a site for the making and remaking of race and racism.

time, it picks up critiques like Tal's and Nguyen's to suggest that present-day diversity advocacy needs to be read as a recentering and redefinition of what counts as hacking—and by extension, who counts as a hacker or technological agent more generally.[33]

A more conventional genealogy of hacking begins, as noted above, with the activities of obsessed technical enthusiasts in the middle to latter decades of the twentieth century that heralded the rise of computing and particularly networked computing. Phones, trains, and mainframes were important early sites for enthusiasts altering technologies, manipulating them to do things they were not built or expected to do (in the Global North).[34] Early networked computing provided a sense of community for its users, and also offered them the sense that this "place" where they convened and communed had its own ethics, politics, and values.[35] In parallel, computer programming itself emerged as a recognized skill with application in employment relations. The growth of this professional category witnessed the "[bifurcation] between free software and the programming proletariat," according to sociologist Tim Jordan. Though both pursuits were based on coding as a practice, different political valences of coding emerged, as practitioners began to distinguish between coding in the service of collectively generated and community-owned projects versus coding for a corporate, for-profit employer or a government employer (in the national interest).[36] All the while, Jordan argues, hacking was tied to consistently unequal social relations that not only uncritically adopted the masculinist biases of computing but amplified them, as the transgressive values and practices of hackers could shade into exaggerated misogyny.[37] Hacking evolved into a self-conscious and increasingly socially prominent (or notorious) community of practice, which often emphasized breaking into (*cracking*) other computers on a network for the purposes of exploring and identifying vulnerabilities. Anthropologist Gabriella Coleman offers a rich description of the lifeworld of a budding hacker, arguing that his identity and commitments emerged through intersubjective experience: "Many hackers did not awaken to a consciousness of

33. This argument is contiguous with yet exceeds the myriad stories that are surfacing in popular culture and scholarship about neglected or forgotten contributions by women and people of color in STEM, such as the 2016 commercial film *Hidden Figures* about four African American women's contributions to the US space program in the early 1960s. See also Rosner 2018.

34. Jordan 2016: 5.

35. Jordan 2016: 5–6.

36. Jordan 2016: 6.

37. Jordan 2016: 6.

their 'hacker nature' in a moment of joyful epiphany but instead acquired it imperceptibly," thoroughly immersed in the cultural and technical facets of hacking alongside others engaged in the same activities.[38]

FLOSS; Hackerspaces and Hacktivism

Simultaneously, free software communities emerged and matured. By the 1990s, free software assemblages could be called a movement, though not necessarily a unified one. STS scholar Christopher Kelty writes, "It was in 1998–99 that geeks came to recognize that they were all doing the same thing and, almost immediately, to argue about why."[39] Free software faced a moment of reckoning with the introduction of the label and similar-yet-different valence of *open source*. In Kelty's words:

> Free Software forked[40] in 1998 when the term *Open Source* suddenly appeared (a term previously used only by the CIA to refer to unclassified sources of intelligence). The two terms resulted in two separate kinds of narratives: the first, regarding Free Software, stretched back into the 1980s, promoting software freedom and resistance to proprietary software "hoarding," as Richard Stallman, the head of the Free Software Foundation, refers to it; the second, regarding Open Source, was associated with the dotcom boom and the evangelism of the libertarian pro-business hacker Eric Raymond, who focused on the economic value and cost savings that Open Source Software represented, including the pragmatic (and polymathic) approach that governed the everyday use of Free Software in some of the largest online start-ups (Amazon, Yahoo!, HotWired, and others all "promoted" Free Software by using it to run their shops).[41]

As this passage illustrates, *free software* and *open source* both do and do not refer to the same thing. The practices that define and unify the mode of production to which these labels refer are sharing source code, conceiving of openness, writing licenses, and coordinating collaborations.[42] At the

38. Coleman 2012a: 30. I use the masculine pronoun here because Coleman does throughout her chapter; she states, "I use 'he,' because most hackers are male" (2012: 25).

39. Kelty 2008: 98.

40. A term in software development that refers to copying code and building it in a new direction, while leaving the original version intact. See chapter 3.

41. Kelty 2008: 99.

42. Kelty 2008: 98.

present juncture, open source has become a much more ascendant label. (Someone I interviewed for this book in 2012 asked me, not at all unkindly, why I was referring to free software in our interview; for him, this was a settled terminology and *free software* was anachronistic). In later writing, Kelty wryly notes, "Free software does not exist. This is sad for me, since I wrote a whole book about it."[43] Though some still hold tight to the free software moniker and attendant political values, it is certainly true that the market- and profit-oriented aspects of these practices have become more authoritative and dominant. At the same time, free software is, by definition, always in a state of becoming; it is a potential, not a fixed value.[44]

A cognate offline phenomenon, with some of the same political ambiguities, is the hackerspace, or hacklab.[45] Hackerspaces port the open-source software ethos to the domain of hardware. Their emblem is the 3D printer, which produces tangible objects whose design is endlessly modifiable. In these spaces, like-minded people come together to hack, learn, socialize and experiment.[46] They are defined by Wikipedia by their devotion to "computers, technology, science, digital art or electronic art."[47] This sets hackerspaces apart from autonomous spaces that are not centered around technology (though technology has had a prominent role in many visions of alternative community since the Appropriate Technology movement of the 1970s[48]). That said, such a space—particularly in a European context—may be located in a former squat, and there is a line of continuity between hacklabs, squatted social centers, and community media activist spaces, including pirate radio and independent journalism[49] (see figure 2.1). Hackerspaces appeared around the turn of the millennium in Europe, picking up steam after 2005, and reaching North America in approximately 2007, via Germany's Chaos Computer Club, Europe's largest association of hackers.[50]

43. Kelty 2013.

44. Kelty 2013.

45. *Hackerspace* is roughly interchangeable with *hacklab, makerspace,* or *hackspace,* though connotations vary. See Davies 2017a, 2017b; Maxigas 2012; Toupin 2015. Makerspaces in particular may share an ethos of democratized fabrication, but they ultimately boil down to leisure and lifestyle, at quite a remove from more radicalized strands of hacking and hacktivism (see Davies 2017a, 2017b).

46. Toupin 2015.

47. Quoted in Toupin 2015.

48. Dunbar-Hester 2014; Pursell 1993; Turner 2006.

49. Maxigas 2012.

50. Maxigas 2012; Toupin 2015.

FIGURE 2.1. A young man peers inside a cabinet-sized radio transmitter, in a warehouse crammed to the rafters with electronics, puppets, and paintings. Visible in this photo in addition to the transmitter are a painted wooden sign that reads "Phoneland" and a PA system for audio. This space is not a hackerspace but a living, art, and social space in an industrial building. It represents a through-line between various kinds of alternative living and technological experiments that precede hackerspaces. Philadelphia, PA, USA, 2004. Volunteer photo, courtesy of the Prometheus Radio Project.

What hackerspaces do have in common is that they are generally not top-down enterprises.[51] They are formed through rhizomatic contact with others in hacking culture, relying on voluntarism for their establishment and maintenance. Most rely on member dues and on financial and in-kind donations to subsist. An especially prominent US hackerspace, Noisebridge in San Francisco, had an operating budget of around $75,000 in 2015, but no staff.[52]

51. Here again, a contrast with makerspaces: some are voluntaristically organized, but others have been placed in high schools, and, at the time of this writing, at least five are run by defense contractor Northrop Grumman (chapter 6).

52. Noisebridge, http://blog.dlenne.eu/mitch-altman/.

This would likely be at the upper end of the range for a US hackerspace.[53] Similarly, initiatives like voluntaristic efforts to teach women to code might receive small grants from foundations. A couple of events I attended had been underwritten by the Python Software Foundation, which covered the cost of pizza and other refreshments for a series of workshops; these were often held in donated space such as workplaces or universities during off hours. And the organization hosting the unconferences for women in open technology that feature in this book was moving toward a more institutionalized part of the spectrum: it was ramping up to having a couple of paid staff and becoming a tax-exempt nonprofit organization before abruptly shuttering in 2015. Its 990 (exempt organization) tax forms indicate its operating budget was approaching half a million dollars in 2015, including individual and corporate donations. This represents the spectrum of institutionalization of the groups in this book; however, most are relatively small, ad hoc, and informal.

Hackers do not possess a monolithic politics. Gabriella Coleman has referred to them en masse as "politically agnostic."[54] But over time, politically motivated hacking, or *hacktivism*, has emerged as a significant stream of hacking. This has included civil disobedience in online settings, ranging from electronic mass protest or virtual sit-ins to launching distributed denial-of-service attacks to paralyze corporate websites. It also includes the creation of tools to protect speech and privacy online, and cracking to obtain information on political adversaries.[55] It is "not strictly the importation of activist techniques into the digital realm. Rather it is the expression of hacker skills [toward political action]."[56]

Women in Computing and Hacking

Many activities in this book represent efforts to foreground feminist principles within the wider hacker or hacktivist milieus. They are intended to extend the reach of what practitioners believe is the emancipatory potential of hacking, expanding hacker culture by making it more inclusive.[57] While they are a more recent intervention into hacking, there are significant

53. Maxigas (2012) states that hackerspaces are more likely to rent, while hacklabs or other spaces in a more explicitly radical tradition are more likely to occupy collectivized or squatted space.

54. Coleman 2012a: 187. Coleman has since chronicled an evolution of hacking politics (2015, 2017).

55. Coleman 2015, 2017; Jordan 2016; see also Renzi 2015.

56. metac0m quoted in Toupin 2015.

57. Toupin 2015.

cultural antecedents that predate the period in this book (approximately 2011 to 2016). The history of women in computing is a complex topic that will receive only cursory treatment here. As historians of computing have shown, women were programmers of electronic computers in their earliest days, assisting the Allied wartime efforts in Great Britain and the United States.[58] Nonetheless, programming was predominantly associated with masculinity within a decade after the war; women's work in computing was effaced,[59] and over the next few decades men flooded the growing computer-related workforce and established the academic field of computer science.[60]

In 1991, MIT computer science researcher Ellen Spertus famously asked, "Why are there so few women in computer science?"[61] By the first decade of the twenty-first century, women's rate of participation in academic computer science had declined even further in the United States. US Department of Education statistics indicate that in 1985, a few years before Spertus's essay, 37 percent of computer science majors were women. In 2009 this number had dropped to 18 percent, and steadily hovered around that percentage during the 2010s at most institutions.[62] This situation is obviously in flux: Carnegie Mellon University boasted that 48 percent of incoming first-year students in computer science were women in 2016, an exceptional highwater mark attained through a concerted effort by university administrators, faculty, and students.[63] As noted in the introduction of this book, the picture in free software was even more grim: the European Union 2006 policy study[64] showed that women's participation was far less than in academic or commercial computer science, coming in at less than 2 percent. Whether or not the gender disparity between FLOSS and other computing was as large as this report indicated, it acquired significance within free software communities, galvanizing attention to women's participation in open-source development.

In parallel with the disparity decried by Spertus, 1987 saw the founding of an electronic mailing list to support women in computing, the Systers

58. See Abbate 2012; Hicks 2017; Light 1999; Misa 2010.

59. Abbate 2012.

60. See Ensmenger 2010b.

61. Spertus 1991.

62. Raja 2014.

63. Carnegie Mellon University, 2016. It is obviously too early to speculate what effect this will have on graduation rates, workplace composition, etc., and Carnegie Mellon as an elite institution has resources other institutions do not. See also Aspray 2016; Margolis and Fisher 2003.

64. Nafus et al. 2006.

list. (See chapter 3 for more on this type of communication.) Systers was formed by "a group of female computing professionals" in response to what they perceived as a men-dominated computing culture, to "allow women in computing fields to communicate with each other in a supportive atmosphere."[65] Women on Systers mostly came from the computing sector in industry and university settings. Similarly, The Grace Hopper Celebration of Women in Computing (GHC) was founded in 1994 by computer scientist Anita Borg. Named for a US Navy rear admiral pioneer of programming, it touts itself as the world's largest gathering of women in computing.[66] FLOSS community attention to gender and other diversity issues thus lagged industry and academic attention to these topics by about a decade.

WE CHEW ON THE ROOTS OF CONTROL AND DOMINATION

But there is another thread of women's participation in computing worth excavating, particularly in the imaginary of who one is when one uses the internet.[67] In the early 1990s, *cyberfeminists* latched onto networked computing as a site of emancipation,[68] creating online spaces that were held to contribute to women's empowerment, participating in identity play, and generally appropriating computing as complementary to, and central to the evolution of, women's identity and experience (figure 2.2). This stood in contrast to an *eco-feminism* that drew on tropes of nature and viewed technology with antipathy, associating it with masculine dominance.[69] Many cyberfeminist thinkers held that digital technologies would facilitate the blurring of boundaries between humans and machines and between male and female, enabling users to assume alternate identities; new technologies were celebrated for

65. Winter and Huff 1996: 32. "Systers" is a play on "sisters" and "sys" as in system administrator.

66. Critics have charged that prominent companies send women workers to the GHC for celebration, while failing to materially support them in workplaces: "Many of the companies in the tech industry who are facing the greatest challenges of gender discrimination are the same ones with a prominent presence at GHC" (Salehi and Tech Workers Coalition 2017).

67. Streeter 2010.

68. "We are a collective body of feminists who . . . chew on the roots of control and domination," read an announcement for a feminist hacking event (Email,—to [Feminist Hacking List], June 25, 2018). Though from 2018, this quote harks back to cyberfeminist collectivity and centers power in its technical engagements and explorations. "Root" here also invokes the root directory in an operating system.

69. Haraway 1991a. Haraway's text is less utopian than many subsequent iterations of these ideas—and cyberfeminism is not a single theory or philosophy.

FIGURE 2.2. "We are the modern cunt" cyberfeminist manifesto by Australian arts collective VNS Matrix, 1991. See Evans 2014a; 2014b. Courtesy of VNS Matrix.

heralding a new relationship between women and machines.[70] Imagining degendering or reconfiguring gender were powerful and liberating experiences for many who pursued these ideas at the crest of the feminist wave.[71] Founders of the 1990s cyberfeminist web portal *Mujeres en Red* in Spain urged users to participate in free software, linking cyberfeminism to FLOSS.[72]

Of course, the past wasn't dead, and it wasn't even past: by early in the next millennium critics and scholars were pointing out that cyberspace was not in fact a space where in-real-life (IRL) social identity got left behind, and that networked computing did not usher in a utopia.[73] From our current vantage point, these may be painfully obvious statements. Notable incidents that have occurred in the past few years serve to render them painfully obvious. Possibly the most notorious to date is 2014's Gamergate, a particularly vicious harassment campaign directed at women gamers,

70. Wajcman 2007: 291.

71. Turkle 1995; see also Stone 1992.

72. Puente 2008: 438.

73. Nakamura 2007.

journalists, scholars, and game developers, culminating in death and rape threats.[74] More recently, a fracas has surrounded Google engineer James Damore's 2017 circulation of a memo in which he opined on "natural" gender differences leading to men's prominence in STEM fields.[75] These events could not have occurred without the ascendance of tech employment, both materially and in the cultural imagination, since the two dot-com booms. In other words, a steady drumbeat—sounded by mass media, national governments, college administrators, and countless others—insisting that employment in STEM fields, computing in particular, is the most viable route to secure future employment invites the entrance of new participants. The uptake of open source within major corporate shops has become widespread. These factors have understandably swelled the ranks of related off-the-clock, voluntaristic pursuits. Thus, while little to none of the action in this book is set in workplaces, the communities I have studied are simultaneously distinct from and yet contiguous with wider tech industry culture. In some ways they are standing on the thin edge of a wedge; what is at stake in their contestations in voluntaristic spaces includes their own communities, the workplace, and wider notions of social good.

Arguably, to be a geek is to assume a subject position in a technologically advanced society where an abundance of gear and surfeit of time (whether one's own leisure time, volunteer labor, time stolen from an employer, or something in between[76]) can be presumed.[77] According to cultural studies scholar Radhika Gajjala, cyberfeminism had a Global North cast, and needed to be adapted to hold promise for those in the Global South; in particular, she questioned whether access to computing could, itself, serve as an equalizing force.[78] Not only does *geek* originate within a largely North American or European cultural context, the export of geek identity can be interpreted as an expression of Global North dominance and a means to bring people in other parts of the world (especially the Global South) into alignment with neoliberal[79] and capitalistic values. It is not a coincidence

74. Classicist Mary Beard provocatively argues that since antiquity, women's voices in the public sphere have often been received with scorn and hostility, unless they are narrowly circumscribed to speak about "women's issues" (2014). See Chess and Shaw 2015.

75. Lecher 2018. As of this writing, the fired Damore was suing Google, claiming that the company discriminated against conservative white men.

76. Söderberg 2008; Turner 2009.

77. Coleman 2012a; Kelty 2008; Söderberg 2008.

78. Gajjala 1999: 618.

79. Streeter 2010.

that some of the values of geek communities, including self-organization and peer production, can be easily ported onto discourses of entrepreneurship and bootstrapping.[80] Some attempts to export geekhood to "Africans,"[81] for example, rightly identify computer capital as a "as a mark of distinction with which to ensure [people's] viability on the job and in the social structure."[82] Yet these attempts often fail to consider the inadequacy of the distributive paradigm[83] as a mode of intervention into systemic inequity. In other words, social power and technical participation are imbricated to such a degree that they may at first glance seem equivalent, but increasing participation in technology is no guarantee of movement into a more empowered social position.

Hacking Hacked?

When technologists seek to advance a technological vision, they are always involved in cultural mediation as much as technological production. In the following chapters, this book will show that diversity advocacy in open technology reveals a complex dialectic between social relations that reinscribe a dominant order and others that begin to challenge, reframe, or rewire this order.[84] National and global economic and labor currents are important contexts for understanding diversity advocacy. So, too, is the underarticulated history of hacking in which peripheral people and practices move to the center of the frame. At present, exhortations to "learn to code" are all but deafening. This book shows why the reply, "learn history and social theory!" is not a snarky rejoinder, but an absolutely essential pointer to the means to effectively grasp the economic, technological, and cultural stakes in contestations over diversity in tech fields, and inequality and pluralism more broadly. Everyone becoming a technologist is not a way out; universalist longings cannot unseat sticky dilemmas of inclusion, belonging, and differential social and economic power. A more nuanced understanding, which can be read through present contestations around diversity in technology fields, is urgently needed.

80. Streeter 2010: 69–70. See also Lindtner 2015 and forthcoming.

81. Roberts 2012. Beltrán (2018) elaborates on these dynamics in a Latin American context, arguing that hacking reorganizes social relations of precarity and uncertainty.

82. Postigo 2003: 600.

83. See Eubanks 2012, chapter 2.

84. Söderberg and Delfanti (2015) write of "hacking hacked" as the dialectic between hacking being recuperated "from below" and the co-optation of (technological) critique by firms and capitalism. See also Ames et al. 2014.

3

To Fork or Not to Fork

HACKING AND INFRASTRUCTURES OF CARE

> The 3D printer will come—now how do we kick people out?
> —INTERVIEW, JULY 24, 2014

When activists engage with technology, social relations—not technological artifacts—are the primary product. This is perhaps an obvious point, but it bears repeating in a cultural moment where technology is a dazzling object of focus. A founder of a "women-only"[1] hackerspace and conference series to promote women* in open technology told me in an interview that initial

1. "Women-only" was here explicitly intended to be inclusive of people who were not cisgendered women, but identified as women in some significant way, which is why I include the asterisk later in this sentence. Over the period of this fieldwork, even older groups that had roots in more essentialist "women-only" spaces came to articulate their intent to include and support people across a spectrum of gender. Newer groups generally affirmed inclusive practices toward women, trans, and gender nonconforming people, though language varied. In other words, though people did often use the term *women*, it tended to be invoked as a political category as opposed to an essentialized or biological one; critique of both biological categories and the social system of gender was assumed, even when people were loose with terms. One person on a feminist hacker electronic mailing list summed up this thinking: "The actual goal is freedom from the undesirable attitudes *which in partriarchy* [*sic*] *are hitched to gender*, and as such [we will] pioneer or navigate the complexity" (Email,—to [Feminist Hacking List], November 1, 2018, emphasis added). This does not mean these negotiations always occurred without conflict; see chapter 7 for more on gender negotiations.

conversations about how the hackerspace would operate were largely centered on governance, not "our wish list of tools."[2] This is pithily expressed in her statement, quoted at the start of the chapter.

Diversity initiatives are aimed at recasting the forms of sociality that determine how technical communities form and proceed. In these formulations, tools such as code, electronics, knitting needles, and Dremels—and even the 3D printer, a fetishized object that signifies open source and *making* like no other[3]—take a backseat. Instead, diversity advocates imagine and create social foundations, rules, and norms that form the basis for association within their communities.[4] This chapter argues that their practices are essentially an extension of the geek practice of argument by technology, an outgrowth of the "rough consensus and running code"[5] ethos that defines free software development. But here the interventions are not technical in the sense of code itself; they are hacks of open-technology communities themselves, which directly flow from practitioners' habit of reflective technical engagement.[6] Diversity advocates are attentive to the layers of sociality that form the architecture of their congregation: they seek to re-engineer the structures that undergird their communities and their practices.

How governance is enacted in communities whose product is free and open-source software is worth special scrutiny. Their ethos has historically been characterized by a commitment to voluntarism and by an ambivalent

2. Chapter 4 is more concerned with the material practices and products of diversity advocacy, while this one focuses on a wider mode of cultural intervention, of which the introduction of feminist hackerspaces is one example.

3. As Lindtner et al. write, "[A] low-cost 3D printer . . . has become a key symbol for an industrial revolution via DIY making" (2014: 5). See also Delfanti and Söderberg 2015.

4. Fox et al. discuss *hacking culture*, which is related. They primarily examine how feminist hackerspaces challenge what counts as hacking, decentering masculine practices and particularly elevating feminine craft (2015). Going further, some advocates of feminist hacking elevate social relations above products. Some write, "Feminist hacking/making does not reify the creation of new artifacts but instead presents itself primarily as a method for encounter and engagement. . . . Feminist hacking/making includes not only teaching underrepresented groups how to do technical things like coding and soldering, but also includes the complex socio-cultural work of bringing technological experts into dialogue with non-technological others. . . . This reorients the stakes of hacking/making beyond the creation or destruction of artifacts but towards the more difficult work of initiating encounters and social relationships" (SSL Nagbot 2016).

5. See discussion in Kelty 2008: 58 and Coleman 2012a: 125–26.

6. In 2005, in the community surrounding the FLOSS operating system Debian, participants submitted a "bug report" to their list for reporting technical problems, requesting that greater attention be paid to gender in their community. As I have argued elsewhere, it is "politically significant that [a developer] turned to a *technical* platform to diagnose a 'social' problem" (Dunbar-Hester and Coleman 2012, emphasis added).

relationship to formal structure, a tension between individualism and collectivity.[7] Free software projects of the 1990s were "experiments in coordination . . . [They were] exemplars of how 'fun,' 'joy,' or interest determine individual participation and how it is possible to maintain and encourage that participation and mutual aid instead of narrowing the focus or eliminating possible routes for participation."[8] While the pursuit of individual "interest" and "joy" does not preclude FLOSS projects from developing an organizational structure,[9] they have traditionally been tolerant of, and even devoted to self-organizing modes where participants determine their own paths through project contribution. It is assumed that individuals will choose to participate on their own terms and largely define those terms.[10]

This has often resulted in naturally emerging hierarchies based on technical skill and reputational capital. Technical expertise is notoriously undemocratic and prone to hierarchy even when participants are committed to democratization, which they often are not, in part due to a historical legacy of engineering as an elite body of knowledge and practice.[11] Technical projects may be especially difficult to run in a radically egalitarian manner, as some people are bound to be more expert than others: egalitarian politics may sit uneasily along unequally distributed expertise.[12] In any event, democratic social relations and loose structure are certainly not equivalent. As Jo Freeman famously argued in 1970 about political organizing within the New Left and the Women's Liberation Movement, structurelessness becomes a way of masking power, not of leveling it. Within open source there has been a strong commitment to meritocratic assertiveness surrounding technical matters. This has been accompanied by a systematic devaluing of the practices within free software production that are less about *coding* (e.g., outreach, community management, documentation). Critics have argued that these factors historically contributed to reduced participation by and recognition of women in open source.[13]

7. Chen 2009; Coleman 2012a: 44.

8. Kelty 2008: 212.

9. See Coleman 2012a, chapter 4 for a discussion of multiple governance modes within one project, Debian.

10. Kreiss et al. 2011: 249.

11. Marvin 1988; Oldenziel 1999.

12. Dunbar-Hester 2014. In her ethnographic study of Debian, Coleman demonstrates that the democratic majoritarian mode of decision-making was explicitly not used to solve technical problems (2012a: 134–35). In any event, democratization is a continuum, not a binary (Eglash and Banks 2014).

13. Lin 2006a; Nafus 2012; Reagle 2013.

In addition, it has been amply documented that in open-source communities the cultural values surrounding technological production were accompanied by a tenacious devotion to free speech and free expression. One consequence of this was that open-source culture largely provided fertile ground for confrontation and hostile speech—including sexist speech—to flourish. Communication studies scholar Joseph Reagle writes, "The anarchic–libertarian ethic requires a significant tolerance for adversariality [*sic*] that may be alienating to some participants. Such participants may actually feel freer to participate under a more structured form of community governance".[14] While norms of civility could vary, the wider milieu tolerated or even bred antagonistic interactions. Of course, uncivil interactions including *flame wars* have historically been part of much online interaction, not only in open source. Nonetheless, observers hold that open source and free culture movements have provided a special "veil of anonymity and freedom" cloaking antisocial and discriminatory speech and behavior by participants.[15]

This background on norms within open-source communities, which is admittedly painted with a broad brush, is necessary to understand the interventions discussed in this chapter. "How do we kick people out?" must be understood as a reaction against norms of openness or structurelessness that have set terms for association with which diversity advocates disagree. Many of diversity advocates' undertakings seek to challenge and recalibrate the norms of the wider culture of voluntaristic technical production. Others provide alternative spaces for hacking and making outside the mainstream. Both strategies are important within diversity advocacy. And both contrast with modes of self-organization that would allow extant terms of association to proceed organically.

In this chapter, I highlight struggles over sociality. I am concerned less with the plugs, cables, circuit boards, thread, or lines of code that geeks hack together than with the templates for *what happens* in the social world of the internet relay chat (IRC), the conference, the meetup, the hackerspace. This chapter argues that these struggles—attempts to establish and embed new standards—also constitute a form of hacking. Current advocacy can be viewed as a moment of breakdown and rebuilding wherein struggles concerning sociality are occurring, such as in the painstaking construction of safer and closed spaces built by diversity advocates, which represent

14. Reagle 2013.

15. Reagle 2013. "Free culture" refers to a movement to broaden the ethos and intellectual property attitudes of FLOSS to other cultural goods.

attempts to build infrastructures of care.[16] It is particularly evident in the challenges diversity advocates raise to the wider world of voluntaristic tech cultures. In the following examples, we see diversity advocates laboring to recast the terms of association in both their wider communities and in spaces they call their own.

Awareness

Changing the Dominant Culture: Visibility and Dialogue in Mainstream Spaces

A common mode of diversity advocacy is simply drawing attention to issues of diversity and representation. Examples are too numerous to track, but here I recount a few instances, from the memorable to the more mundane. All of them occurred after the 2006 EU policy report showing that women's participation in open source was very low, as described in preceding chapters.[17] As far as I know, they were not coordinated, though communication and contagion at the network level can be assumed. These episodes should be interpreted as interventions directed at the wider open-technology culture as well as the culture more broadly.[18]

16. This is related to repair work, in the sense suggested by Steve Jackson (2014). Lilly Nguyen brings together notions of infrastructure and care when she writes, "infrastructural actions are characterized by ongoing attention, care, and hence must be performatively constituted on a regular basis" (2016: 638). In describing terms of community association as "infrastructure," I do not mean to imply that other infrastructures do not contain social elements; infrastructures are already always sociotechnical (see Hughes 1987; Star 1999; see also Carse 2016; Edwards 2003). In any event, my use of "infrastructure" here is a loose, metaphorical usage, *not* a defensible analytical category. I merely wish to draw attention to how in my empirical sites there is a usually unremarked-upon layer that structures practice, which has now become visible, remarked upon, and reworked in a moment of breakdown, with the aim of re-standardizing association along new lines; "infrastructure" thus fits pretty well as an orienting concept. Precedent for highlighting social infrastructure can be found in a discussion by Langdon Winner: "The important task becomes . . . not that of studying the 'effects' and 'impacts' of technical change, but one of evaluating the material and social infrastructures specific technologies create for our life's activity" (Winner 1986: 55). Benner refers to "soft" infrastructure, but this seems worse than "social," not least because of the gendered language (2003). Irani et al. also use the term "social infrastructure," this time meaning something closer to "social capital" (2010). Klinenberg calls for renewed attention to "social infrastructure," meaning spaces for public and civic life, in his *Palaces for the People* (2018). And Facebook's Mark Zuckerberg used the term "social infrastructure" in a 2017 blog post, but arguably he was trying to distance himself from earlier language where he described Facebook as a "social utility" because of the regulatory treatment that might invite (https://www.facebook.com/notes/mark-zuckerberg/building-global-community/10154544292806634/, February 16, 2017).

17. Nafus et al. 2006.

18. They also might be thought of as agitation by subaltern counterpublics, following Nancy Fraser (1990).

In July 2006, a few attendees of the Hackers on Planet Earth (HOPE) Conference in New York City offered for sale homemade, silk-screened t-shirts that riffed on a recent gaffe by Senator Ted Stevens (R-AK). Stevens, a senator charged with regulating the internet, had recently remarked, "The Internet is not something you just dump something on. It's not a truck. It's a series of tubes."[19] This comment was widely circulated online and roundly mocked by many who insisted that Stevens understood neither the technical aspects nor the principles of the regulation he was crafting, which concerned *net neutrality* and whether internet service providers should be barred from giving delivery priority to favored content.[20]

The activists hawking t-shirts at the HOPE conference swiped not only at Stevens's attack on net neutrality. They also lobbed a separate critique at their own community: their t-shirt featured Stevens's quote "The Internet is a series of tubes" as the caption for an anatomical representation of the female reproductive system (figure 3.1).

One mischievous woman told me that she had refused to sell shirts to men unless they first said out loud to her, "uterus" or "fallopian tubes." In other words, these activists creatively and humorously challenged the assumption that a technical domain such as the internet is a masculine one.[21] Significantly, they did so at a conference for computer hackers, an event dominated by men speakers and audience members. (Participants estimated the ratio of men to women as 40 to 1, though there is no way of verifying this officially.) In this context, asking men to say out loud words related to female reproductive organs before letting them buy t-shirts was a flag-planting gesture. It also provided fodder for storytelling after the fact (as it was relayed to me), emphasizing both the scarcity of women in spaces like hacker conferences and the fact that this scarcity was not proceeding unchallenged.

And this episode is only one minor, fleeting example. Contestations surrounding who participates in technology production abound in recent years. In 2007, to commemorate International Women's Day, feminist techies based in Europe coordinated a virtual march through IRC channels. They adopted handles associated with women in technology, including Ada Lovelace and

19. *Wired* 2006.

20. The mockery of Stevens appeared as widely as Jon Stewart's *Daily Show* on the Comedy Central cable network See for example, *The Daily Show* "Net Neutrality Act" segment, July 19, 2006.

21. Cleverly extending the metaphor into reproductive politics and women's right to choice, the back of the t-shirt read "Senator Stevens, don't tie our tubes!" See Rentschler and Thrift 2015 on feminist memes.

FIGURE 3.1. Activist T-shirt, "The Internet: A Series of Tubes," 2006. Courtesy of Steph Alarcón.

Grace Hopper, and other feminist figures from history, literature, and pop culture, including experimental novelist Kathy Acker, musician-performer Peaches, and Victorian writer George Eliot, and "marched" through IRC spouting feminist slogans. An excerpt:

PEACHES: When men are oppressed, it's a tragedy. When women are oppressed, it's a tradition.

GRACEHOPPER: It's better to act on a good idea than to ask for permission to implement one.

CHARLOTTEPGILMAN: When the mother of the race is free, we shall have a better world

* [s—] ([s@ IP address]) has joined #backchat

TUX: hey sister welcome to #backchat!
CHARLOTTEPGILMAN: happy iwd [International Women's Day] 2007!
SIMONEDEBEAUVOIR: Well-behaved women seldom make history
* [M—] ([m@ IP address]) has left #backchat
GRACEHOPPER: Bread and Roses[22]

According to the organizers of the march, marchers were kicked out of a number of IRC channels because other users thought they were *bots* due to the coordinated nature of their appearance: "Naturally we were deftly kicked and banned from most servers as a result of our actions. One set of tech operators apologised and lifted the ban when they realised we weren't bots: they found us so co-ordinated they couldn't believe it to be otherwise."[23] They make this claim with obvious relish, because it signifies the marchers' effectiveness at creating a spectacle. It constitutes storytelling about the marchers' disruptive feminist and feminine presence in a space where masculinity tended to reign without comment. Moreover, the marchers' claim that they were assumed to be bots rather than actual live women users works to establish the ostensible strangeness of feminine presence in such online spaces.

Another approach is to have presentations and discussions that raise awareness of diversity issues in mainstream conferences—those centered around hacking or open technology in general, as opposed to diversity issues. Shortly after the release of the 2006 EU report, advocates for diversity had collated the research and began presenting it to audiences at software conferences. For example, a computer scientist who cofounded Debian Women presented data drawn from the report at a number of conferences. Such presentations worked to establish the perceived problem of diversity in FLOSS and justified attention to this issue at a time when its validity could not be taken for granted in the wider community. Her presentations included quantitative metrics of FLOSS community members' experiences and observations about their communities (figure 3.2).

And she also made gentle gibes about the presence and visibility of men relative to women. In another slide, the researcher has inserted her own face covered with a knitted beard alongside some of the iconic (male, bearded, and mustachioed) faces of open-source programmers, with the

22. Genderchangers.org n.d.
23. Genderchangers.org n.d.

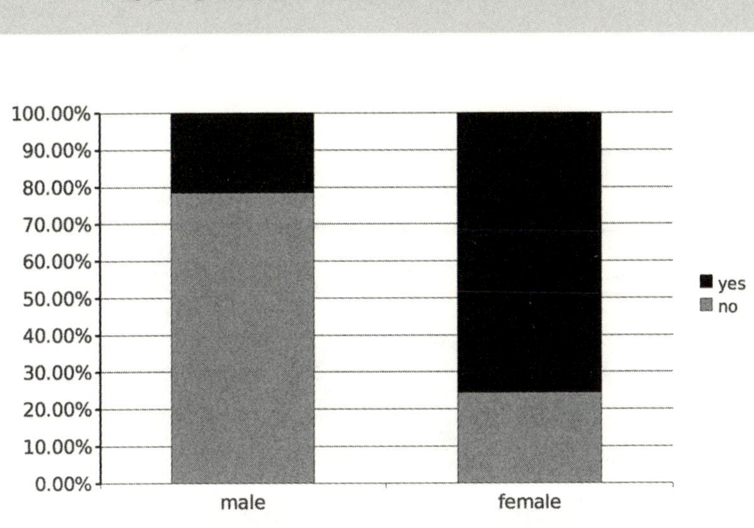

FIGURE 3.2. Slide, FLOSS participants' observation of discriminatory behavior. Wallach 2010, drawn from Nafus et al. 2006. Courtesy of Hanna Wallach.

humorous caption, "Do I Have to Have a Beard?" (figure 3.3). Once again, this image establishes the strangeness of femininity in a sea of masculine presence. Notably, the masculinity displayed here is linked to a nonmanagerial, countercultural presentation of self in the form of longer, unkempt hair, facial hair, and t-shirts and casual clothes, belying a high degree of expertise.

These interventions into the mainstream spaces of open-source communities and hackerdom continued for many years, sometimes taking the form of data presentations, other times as reflection, consciousness-raising, and dialogue. At a subsequent HOPE conference in 2012, also in New York City, two leaders of a feminist hackerspace in Philadelphia ran an exercise, "Hacking the Gender Gap," for an audience of around twenty-five people, which I attended. They asked everyone in the audience to write on Post-It notes positive and negative experiences with technology, their age at the time, and their gender. They then took the notes and arranged them to illustrate patterns, which they shared with the audience along with the specific stories the audience had generated. Some of the patterns that emerged were that parents, especially fathers gave early positive experiences with technology, and that having early negative experiences could affect people for a long

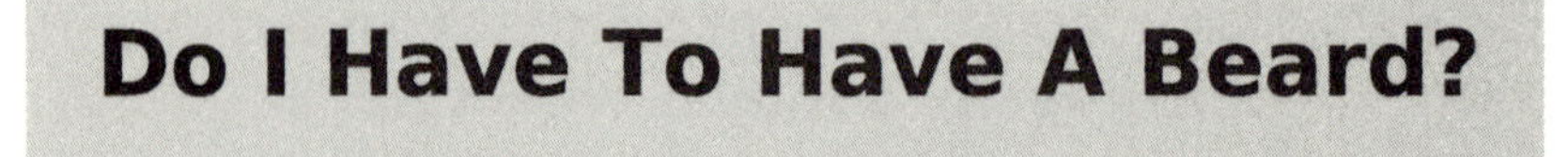

FIGURE 3.3. "Do I Have to Have a Beard?" slide, a playful challenge to hegemonic masculinity in hacking culture. The people in the slide include the speaker in the lower right as well as Richard Stallman, author of the GNU General Public License; Alan Cox, Linux kernel programmer; Larry Wall, Usenet client author and creator of the Perl language; and Henk Penning, prominent contributor to Apache web server software. Wallach 2010. Courtesy of Hanna Wallach.

time. They showed that teachers could be a source of positive mentorship or of direct and potent discouragement.[24]

They then opened the floor for discussion. One younger man in his twenties, with visible piercings and primitivist tattoos, spoke up immediately and said it had been hard for him to conceptualize "stories about tech" from his life because his "whole life was tech"—he had had computers from the earliest moments of his upbringing. (Studies of adults in free software have showed that girls and boys encountered computers at different ages, with girls' use and computer ownership starting later.[25]) Two older men spoke up in response to this comment, one saying he had learned to program on a legal

24. Fieldnotes, July 14, 2012, New York, NY.
25. Wallach 2010.

pad and another, by using the phone to tell a person on the other end to enter his code and run it on a computer. I noticed that men were chiming in more than women.[26] I was not sure whether this was in part because the organizers were explicitly hoping to have men reflect on their own gendered experiences with computing (reflection which was possibly somewhat uncommon for them), or whether this was an unconscious replication of the dynamics the workshop meant to address: men felt more comfortable speaking up and were not necessarily aware that they were dominating the conversation. It may have been some of both. I noted that while the room did not seem tense, there was a bit of a charge to the discussion, and participants were animated, and some perhaps a bit defensive.

People also discussed how language, in such formulations as "the tech guy" and the generic use of *he* to mean *technologist* created what someone termed a "slow-moving psychological barrier, the feeling that you [as a woman] don't belong here."[27] Clara, a workshop organizer, said that the personality of a hackerspace or project is formed by the tone and interactions that happen there. She said, "If you want your project or space to have the toughest people who can hack something, you will get those people, but you won't get other people. It's a smaller and less diverse demographic [who will show up with that dynamic]."[28] The workshops being run in the Python community for "women and their friends" came up numerous times in discussion and in examples of women's positive experiences.

The next day, in a much larger session (about two hundred people in attendance), HOPE attendees continued to discuss inclusion and diversity in a panel called "Hacking the Spaces." This titled referenced a 2009 pamphlet of the same name cowritten by one of the HOPE panelists, Johannes Grenzfurthner, an Austrian artist in his thirties. Grenzfurthner gave a history of hackerspaces, locating their origins in the failed utopian projects of hippies and the 1960s and 1970s counterculture, which then morphed into later countercultural formations of punk subculture and the squatters' movement of the 1980s which, he said, birthed DIY.[29] He said that "one aspect of early hackerspaces was, we want to live somewhere that no one takes care of [like

26. Fieldnotes, July 14, 2012, New York, NY.

27. Fieldnotes, July 14, 2012, New York, NY.

28. Fieldnotes, July 14, 2012, New York, NY.

29. A more historicized understanding of DIY would include its earlier origins as a masculine consumer project of home improvement in the postwar period (Gelber 1997). But DIY was rediscovered and refashioned in the political and technical imaginaries of squatter and punk subcultural scenes in the 1980s (Waksman 2004; Dunbar-Hester 2014).

an abandoned building or warehouse], so we need tools. . . . There would not be as many hackerspaces nowadays without . . . a long countercultural tradition."[30] He condemned present-day hackerspaces for forgetting the history of the movement, reducing political potential to "I just want to make something with an Arduino." (The latter is a small circuitboard, the function of which can be set and controlled using open-source software; notably the board design is also open.[31]) He chided present-day hackerspace members, particularly in the US, for possessing a limited understanding and imagination of hacking, and for whom "working with an Arduino" was more of a hobbyist pursuit, decoupled from the political motivations for hacking.

His fellow panelist, also a white man in his thirties, who had cofounded a hackerspace in Los Angeles, said that he held hackerspaces to a higher standard than makerspaces. For him, in a hackerspace, community was the most important element, and he said his space would turn away people who just wanted to use their tools but did not want to come to meetings and be actively involved in the ethos of the space. By contrast, he said that it was OK for "someone who just needs to use a laser cutter" to approach a makerspace—but he was not interested in participating in such spaces.[32]

Continuing to dissect the failed promise of hackerspaces, Grenzfurthner raised issues of social inclusion. He claimed that in spite of their self-promotion as being some of the most open spaces, hackerspaces were "usually white, male, heterosexist." He commented that, "Even in Germany, where there is a longer tradition of hackerspaces than in the US, it's hard to find Turkish members in German hackerspaces." Speaking of a neighborhood of Montréal, Canada where a hackerspace was located, he said you might find in an area "where predominantly black people live, you might find a hackerspace that doesn't have a single black member. . . . We're the gentrifying nerds, but we're not trying to talk to the people who live next to us. That's kind of bad. Same problem in Germany, even in Turkish neighborhoods, not a single Turkish member."[33] He attributed this to the "class-based structures of Western societies" that contribute to urban social geographies in which hacking groups might occupy storefronts or industrial spaces in the same areas as cities' poorer residents. While not blaming hackers for the existence of these stratifications, he argued that the present practices

30. Fieldnotes, July 15, 2012, New York, NY.

31. See Powell 2012 for a discussion of the history of open hardware.

32. Fieldnotes, July 15, 2012, New York, NY.

33. Fieldnotes, July 15, 2012, New York, NY.

of hackerspace communities were not up to undertaking dialogue with or being truly welcoming to neighbors or community members different from themselves, and he saw substantial room for improvement.

Opening this presentation up for discussion, the panelists invited a handful of audience members to join them at microphones on the stage for greater variety in the discussion on stage and with the audience. Two tensions emerged in the discussion that are worth comment here, for they illuminate some challenges for hackerspaces and other groups committed to openness. First, Europeans touched on various hackerspaces in Europe, including a situation in Bielefeld, Germany, a relatively small city with multiple hackerspaces that had formed when hackerspaces had experienced disagreements and divisions. Joe, a white man in his thirties from Philadelphia said that he was strongly in favor of forking groups when differences too great to overcome emerged among members. (*Forking* is a term borrowed from software development, meaning to take the existing code, copy it, and build it in a new direction, while leaving the original code in place for subsequent development as well. This is a prevalent practice in open-source development.[34]) He said, "I left a [hacker]space over frustration, I was trying to drag people into caring [about diversity and inclusion]. I felt that forking the group was better than changing the group [especially when other members resisted that change]." Grenzfurthner vehemently challenged this idea, saying, "From a political standpoint that's the worst thing you can do, factionalizing marginal groups." He felt that it was vastly preferable to have ongoing dialogue, even uncomfortable dialogue, within one large group. Joe from Philadelphia disagreed, arguing that it was "better to have three groups in town who are friendly but don't work on the same projects."[35] As will be discussed further below, many diversity advocates have opted for forking groups—starting over in new spaces with more like-minded people. The radical openness espoused by Grenzfurthner has, in practice, been perceived as difficult and burdensome, even leading to hurtful interactions. In any event, the discussion indicates that this issue is a challenging one for groups that share a normative belief in openness.

Panelists also discussed what they called elitism in hacker culture. Examples included geek/nerd culture and self-expression, where it is usual for people to wear clever t-shirts that relate to subcultural, insider knowledge, some of which is technical in nature and some of which is not technical

interesting discussion

34. See Tkacz 2014 for a sustained discussion of the politics of openness and forking.
35. Fieldnotes, July 15, 2012, New York, NY.

per se but is arcane, related to comics, video games, and similar. Making an unsubtle parallel to displays of masculinity,[36] Grenzfurthner likened hacker identity performance to a person saying, "Here is my dick hanging out from my chest [on my t-shirt]. If you don't laugh at the joke, you are beneath [me]. Could be ironic, but it is still my dick." The implication was that geek humor and subculture set up a hurdle, which only like-minded people also steeped in these topics would clear; subcultural references (to say nothing of technical expertise) set up a barrier to entry. Panelists acknowledged that there was an incongruity in construing hackerspaces as open to everyone, including people of different genders, cultural backgrounds, and socioeconomic statuses when at the same time, as Grenzfurthner put it, "isn't a key part of being in a hackerspace being an elitist?"[37]

This communicative repertoire I have described was intended to increase awareness about diversity issues in open-technology communities. It could be characterized as agitation by a subgroup, directed toward a wider open-technology public. It established feminine presence, and mounted critique in the form of consciousness-raising, reflection, and discussion. In all of these actions, diversity advocates attempted to change the terms of association that undergirded their communities.

Changing the Dominant Culture: Codes of Conduct

In a 2013 blog post, a woman wrote, "When trying to explain how hostile an environment the geek world can be, I'd tell people, 'I've been attending cons [conferences] of various types since I was thirteen, and I have never, not once, been to a con where I wasn't harassed.' "[38] The author went on to describe attending a particular open-source software conference for the first time, including her apprehension beforehand at entering a new community and her pleasant relief that the new conference experience was one where she "felt safe the entire time," where "nobody made inappropriate sexual comments, touched me without my permission, or took creepy pictures of me."[39] In the post, she argued in favor of conferences adopting codes of conduct, as this one had. While she did not attribute her positive experience

36. The version of masculinity prevalent in geek or nerd cultures is a nonhegemonic masculinity, but a form of masculinity nonetheless. See Bell 2013; Dunbar-Hester 2014, chapter 3; see also Connell 2005.

37. Fieldnotes, July 15, 2012, New York, NY.

38. Geekfeminism.org 2013.

39. Geekfeminism.org 2013.

solely to the code of conduct, she said it was useful for laying out rules and accountability for everyone in attendance; it also broadcast that the organizers of the event cared about the experiences attendees would have. She said that codes of conduct "tell me what the community's standards are. They also tell *everyone else* what the community's standards are, which is perhaps their biggest effect: telling people what behaviors are unacceptable can prevent problems from occurring at all."[40]

During the period discussed in this book, agitation for codes of conduct was a major phenomenon within open-technology circles. Diversity advocates insisted on the adoption of clear rules for comportment in various spaces where open-technology communities congregate (which spanned both online and offline spaces: IRC channels, mailing lists, user group meetings). Codes of conduct (CoCs) were controversial in various ways,[41] with some people wondering whether they provided a false sense of security and urging greater specificity, including concrete steps for carrying out and enforcing policies; as one commentator wrote, "We cannot even be sure that antiharassment policies are anything other than security theater."[42]

Substantially more criticism held that codes of conduct were unnecessary or detracted from the real focus of open-source projects, or represented moral policing that critics found unsavory. Arguments about these topics are entirely too many to enumerate, but a fairly typical line of discussion occurred when GitHub proposed adding a code of conduct in 2015. (GitHub is a repository that hosts code for many projects and is thus an umbrella site for open source, which is commonly visited by people from a huge variety of projects.) When it added a code, the open-source community exploded with discussion, such as the below comment thread on the site *Hacker News*:

> COMMENTER A: Code of Conduct defines problems and could turn anything funny, innocent, etc. into an excuse to ruin someone's life. 9/10

40. Geekfeminism.org n.d. Emphasis in original.

41. This book is not symmetrical in describing diversity advocacy versus opposition to diversity advocacy; its emphasis is a range of advocacy activities and their consequences. However, the conversations surrounding CoCs, including opposition to CoCs, are important because they reveal common patterns of how these issues are understood in FLOSS. These contours help to establish why diversity advocates have made some of the choices they have made about how to challenge and rebuild social infrastructure. I do not speculate on the actual personal or social identities of the individual commenters in this vignette, which are not especially important for the analysis. The main point is that these debates are charged and that people are staking positions that are tied to a patterned legacy of open source participation and visibility.

42. Byfield 2015; see also Byfield 2013.

times it's someone who doesn't program ([Name X or Name Y[43]]) who will be attacking someone who actually contributes.

COMMENTER B: You seem to be continuing to operate under the assumption that only people who can code, are capable of contributing to a project. There are so many things that can be done to contribute to a project, that don't involve writing a single line of code. Testing, evangelism, documentation, support, etc, etc. Why should attempting to improve the culture of a project, be any different?[44]

Many who opposed CoCs understood the conflict to hinge on whether an open-tech space should be more like a "clubhouse,"[45] with a freewheeling, voluntaristic ethos that welcomes free expression and a "come as you are" mentality, or more like a workplace, with a list of rules handed down by a human resources (HR) department and the attendant threat of punishment for perceived transgressions (as indicated in the line about "something innocent or funny being an excuse to ruin someone's life"). They defended the former and pushed back against the latter.

First, and most saliently, in the comment thread above, there is conflict over gender. Commenter A invokes the full names of two women (here referred to as Name X and Name Y), both of whom had been subjects and objects of extreme controversy. This commenter also suggests that these people, not being *programmers*, are less valuable to the community, and indeed configures them as distractions, detractions from and even threats to the real work of programming. The underrepresentation of women in open source overall has been found to be more nuanced; even before the 2006 Nafus et al. study, scholars and practitioners recognized that while women were *particularly absent* in the work of coding, they were more present *and yet overlooked* in noncoding work, such as NGO work that supported the development of FLOSS, documentation translation, book editing, teaching, and tutoring.[46] So we should read this argument over "programming" (as invoked by Commenter A) versus "testing, evangelism, documentation, support, and improving the culture of the project" (as Commenter B writes)

43. Here this commenter refers to two people, both women, by name. I omit their names here because they have been at the center of controversy and subjected to harassment; their individual identities are not relevant to this analysis. But the reader should know that they are women whose names would be familiar and probably controversial to others in this discourse community.

44. Github 2015.

45. Margolis and Fisher 2003.

46. Lin 2006b: 1287.

as suffused by, if not reducible to, a conflict over the roles, presence, and relative value of men versus women in open source. (The Donglegate episode later in this chapter is, among other things, an eruption of these issues.)

Communal tech spaces online or off have tended to retain a clubhouse, "come as you are" ethos. The attachment to this sanctuary is quite strong, not least because it has been a space where geeky men could freely congregate out of sight of the enforcers of hegemonic masculinity, not only of women. Open-technology cultures were thus spaces of refuge for their participants. Even when gender is not mentioned outright, it lurks in the background. It is no coincidence that the clubhouses of open technology have been places for (nonhegemonic) masculinity to reign, free from the potentially disciplining gaze of HR. (It should be noted that this freedom from women's interference is about preventing voluntaristic spaces from being made to feel not only like the workplace, but also like the domestic space, as historically men have been accustomed to carving out space in garages, basements, and ham shacks, where they are free to code and tinker without disrupting the domains of the home that have been constructed as feminine, like the kitchen, living room, and marital bedroom.)

As another commenter noted in a separate discussion about the CoC proposed by GitHub, "Back in the days [when] open source was the nerd elite [it was] super unwelcoming towards anyone that didn't strive to excel at what they did. Not the most friendly environment, but it certainly pushed people to do their best in order to gain acceptance."[47] This comment is rich in its valorization of the "nerd elite" and "striving for excellence"; it also echoes the discussion in the "Hacking the Spaces" panel about the difficulty of decoupling technical exultation and expertise from elitism. And it hints at how this elitism may breed competitive dynamics in collaborative projects. People who could demonstrate technical chops and value for a project were welcomed and celebrated; people whose abilities were still in development would ultimately be welcome if they could learn, but they would possibly need to tolerate hazing along the way. (Some older men in the "Hacking the Gender Gap" workshop described above mentioned being hazed when they learned to program.) Though gender is not explicitly mentioned by

47. Myrpl 2015. The Red Pill, hosted by the wide-ranging forum/community Reddit, is a men's subcultural forum. Its description is "Discussion of sexual strategy in a culture increasingly lacking a positive identity for men" (see Marwick 2013; Banet-Weiser and Miltner 2016). This forum is one where posters feel emboldened to express views on gender and open source (among many other topics) that might cause greater controversy in other settings, but there is no reason to assume these particular quotes represent fringe viewpoints.

Commenter A, to be comfortable with competitive displays and feeling unwelcome along a path to greater skill acquisition can mean being comfortable with masculine interpersonal styles, which are especially common in technical realms where virtuosity is prized.[48] This does not mean that all women are put off by such interactions, nor that all men are comfortable with them, of course.

The notion of a nerd elite also reinforces this as a space outside the mainstream. Originating as terms of insult, *nerd* and *geek* have been reclaimed as self-aware and even fond appellations within these communities. The *nerd elite* refers to a community of people who have something in common with each other, but are set outside of and have likely been rejected by and chosen to reject a wider social milieu that includes hegemonic masculinity. The term would almost be a contradiction in terms, except for the solidarity it offers among fellow outsiders. Thus, a sentiment that a hackerspace is "our place" is a strong one and attachment to it is dear. The community is encoded with certain assumptions about who belongs there and what behavior is anticipated, tolerated, and celebrated. The commenter discussing the "nerd elite" also offered the provocative opinion that "if coding wasn't one of the higher paying careers out there these days with relatively low entry barriers for those willing to learn, in an otherwise shitty job market, with startups being perceived as cool as well, there sure as hell wouldn't be any women looking to get into it (except asians [*sic*], which were always pushed towards STEM careers by parents)." This underscores several notions that are key to understanding some of the churn, conflict, and outcry around diversity issues. First, the commenter notes that the coding "clubhouse" was not only historically unwelcoming, but actually undesirable to outsiders; it is implied that this was a mutually satisfactory state of affairs for both those inside and outside the clubhouse.[49] Second, women are the most obvious class of outsiders. Third, the commenter invokes race as well, implying that cultural and familial backgrounds of Asian women might have predisposed them to looking twice at programming even despite their outsider status as women

48. See Douglas 1987 for historical origins of technical masculinity in ham radio; see Dunbar-Hester 2014, chapter 3 for a discussion of these dynamics in amateur technologist groups.

49. As another commenter on the Red Pill page wrote, "I'm in my mid 30s now, but I was made fun of [in] high school for being into computers. Now that it's seen as socially acceptable, you want me to take my time to teach you, for free, under the guise of being 'welcoming'? No thanks" (Myrpl 2015).

or girls.[50] The dominant cultural category of whiteness is not named, but in total this comment can be read as an indicator of white masculinity being hegemonic within open-source communities.[51] Of course, there is a deep irony in the fact that those who found respite from a world in which they might feel excluded or persecuted would reproduce some of these dynamics in their own sanctuaries.

People also pressed for a (somewhat paradoxical) no code of conduct policy for online spaces in the heat of ferment around GitHub's CoC. In late 2015, someone wrote a script for a No Code of Conduct (NCoC) that could be added to the root directory of a project and described the NCoC concept in a Q&A page: "What is NCoC." The page is more a defense of various traditional norms and values of open source than an exhortation to conduct open-source projects without rules of any kind. In particular, it emphasizes that people's political or social identity should be irrelevant in open source: "We don't care if you're liberal or conservative, black or white, straight or gay, or anything in between! In fact, we won't bring it up, or ask. We simply do not care."[52] This strongly echoes the Declaration of Independence of Cyberspace penned by John Perry Barlow in 1996, which reads, "We are creating a world that all may enter without privilege or prejudice accorded by race, economic power, military force, or station of birth. We are creating a world where anyone, anywhere may express his or her beliefs, no matter how singular." The author of "What is NCoC" furthers this parallel in the following statement: "If you limit yourself to participating in things that only have codes of conduct, then you're limiting yourself as a human

50. Lisa Nakamura offers thoughtful analysis of Asian Americans' status online. She notes that English-speaking Asians in the US have historically been over-represented in studies of US internet use patterns while non-English-speaking Asians are rendered nearly invisible. Thus non-English-speaking Asians in the US are a digital minority whose racial formation and public perception are that of a digital majority (2007: 273). Of course, there is a lot to parse vis-à-vis the remarks made by the commenter: we do not know where the commenter is located, we do not know which Asians are being referred to, and coding knowledge or open-source participation is different from the online activity represented by the studies Nakamura discusses. Nonetheless, the point stands that Asian people are in a special position in terms of assumed techno-proficiency, which contributes to technology being a domain where racialized ideas form (for example, about "wired" workers and consumers). This assumed special position not only elevates certain Asians symbolically, it also serves to render invisible much work performed by Asian laborers to support the IT industries that is not high-status, especially offshore work.

51. The race of white people is particularly unmarked online and underexamined in internet-studies scholarship (Daniels 2013).

52. Github.com n.d.

being on this marvelous place we call the internet. . . . The internet is a big place, you should prepare yourself to deal with it."[53]

The author does note that noninternet communications "understandably are different. Often sane communities will have a CoC that is unique to their physical location, and/or event. . . . We are talking about the internet here, not physical meet ups, and that requires *we do not care about the unimportant parts of how people talk*."[54] In other words, the author defends *online* projects as spaces that transcend and supersede how social identity is lived offline, a fantasy of a perfect liberal sphere. Though the author allows for differences in social identity having the potential to affect how people relate or communicate offline, the ideal is for none of this to matter in the realm of open source and online communication. Implicitly, none of this *should* matter at all (identity is part of "the unimportant parts of how people talk," unlike code), but the author does not go so far as to claim this is actually true in real life. Instead, "the marvelous place we call the internet" is a place to experiment with and then practice leaving behind social difference. Thus it is perhaps unsurprising that the suggestion to add a bunch of rules to open-tech spaces has been met with dismay and even hostility, as it undercuts the fantasy that the internet is a space for social leveling. This fantasy has been borne out in some ways for the nerd elite, who found community in online association. Arguably, their experience is one factor in their resistance to acknowledging and accommodating other kinds of experience and difference.

Advocates for codes of conduct, meanwhile, see the issues rather differently. They are often people who have not experienced open-tech communities as spaces where one could join (online or offline) on an equal footing with other members regardless of social identity. This exchange, also from the 2015 *Hacker News* thread, begins with a familiar line of argumentation against CoCs:

> COMMENTER C: The only thing this code of conduct does beside state the obvious (behave like adults) is the implied distrust expressed by introducing it in the first place: that without being nannied into proper behaviour, we would all act like vile animals. Generally I find such nannying deeply offensive.[55]

hmm

53. Github.com n.d.

54. Github.com n.d., emphasis added.

55. Github.com 2015. Once again, "nannying" is an overt reference to (and swipe at) feminine gender. It also draws attention to the poster's belief in self-governance, i.e., possibly libertarian politics and distaste for "the nanny state." Thanks to UCLA Information Studies colloquium participants for drawing this out.

The commenter holds that codes of conduct are beyond unnecessary; indeed, they are presented as belittling and divisive. Another commenter then chimes in in favor of CoCs:

> Commenter D: Have you guys missed the numerous controversial issues that consistently come up around this? I really don't understand the viewpoint that these guys are just coming out with a rule book apropos of nothing except their desire to nanny and police people.
>
> Is it possible that you're a part of the class that, yes, doesn't experience the negative aspects of the community, and thus has no empathy for those that do? . . .
>
> Personally I find that attitude an excuse to be callous and lazy. Actually taking the time to understand other people and their issues, truly trying to get outside of your own biases and experiences and put yourself in their shoes, is more impressive to me.[56]

Commenter D defends the idea that even if many people have never experienced troublesome behavior in open source, reports from people who claim they have are not to be dismissed. It also leaves open the possibility that one person's merely "not warm and helpful"[57] experience in a given community might be experienced as extremely alienating to another person for a variety of reasons. This commenter elevates the potential for differences in how participants are located within the community—there are multiple "classes" within the wider open-technology community—and holds that CoCs may do some work to raise awareness and empathy levels about these differing backgrounds and experiences. Commenter C, meanwhile, argues for personal responsibility, equality among adults, and, implicitly, the right to self-determine and self-organize, without the introduction of "distrust."

Though patently contentious, codes of conduct have gained momentum as one form of intervention into open-technology communities. Plainly, there is a spectrum of belief and practice concerning how to characterize, let alone solve, the problems these commenters identify and tussle over. Even when people are *not* arguing for free speech absolutism (including the right to be as "vile" as possible), how to open up the ranks of open technology to more people presents formidable challenges as its champions reimagine the terms for association.

56. Github.com 2015.
57. Github.com n.d.

Separate Spaces: Hackerspaces and Unconferences

With the recognition that codes of conduct and other interventions into the wider culture and spaces of open technology were insufficient by themselves, diversity advocates have also moved to create spaces outside the mainstream. This section outlines a handful of examples of the creation of infrastructures of care to support communities that are distinct from (but may overlap with) wider open-technology cultures. These are examples of smaller, forked communities. Many advocates pursue both mainstream intervention and separate spaces simultaneously, recognizing that each has utility and that it is not an either/or situation. As Liane, a forty-something unconference organizer, said in an interview, "We change the culture by starting our own space. It's not a binary of work within or fit into [the culture], or just be separate."[58]

The most prominent iteration of these separate spaces is the proliferation of feminist hackerspaces. According to activist-scholar Sophie Toupin, "Feminist hackerspaces can be understood as safer spaces where a set of common values is foregrounded by its members. Feminist hackerspaces are not an end in themselves, but rather a means to address some of the felt shortcomings of hacktivism and to possibly help create a stronger feminist hacker (counter) culture."[59] *Safer spaces* rhetoric indicates that feminist hackerspaces' founders emphasize providing welcoming environments where threatening or exclusionary behavior is not tolerated. A symbolic safer space environment within a feminist art studio in Montréal can be seen in figure 3.4, in which a tent has been constructed from hanging sheets. During a feminist hacker event, a workshop on consent occurred within the tent. Its fabric walls were only flimsy reinforcements, but to enter them was to enter a space where the architecture marked separation from the world at large and encouraged communication and reflection in a supportive environment. These commitments to safer spaces in some ways run counter to certain interpretations and norms of openness upon which open source rests,[60] but organizers unapologetically come down in favor of enacting openness in a critical and reflective way, which is less absolutist and more conducive

58. Interview, July 24, 2014, San Francisco, CA. See Christian 2018 for a study of these dynamics in the entertainment industry.

59. Toupin 2015; see also Davies 2017b.

60. Reagle 2013.

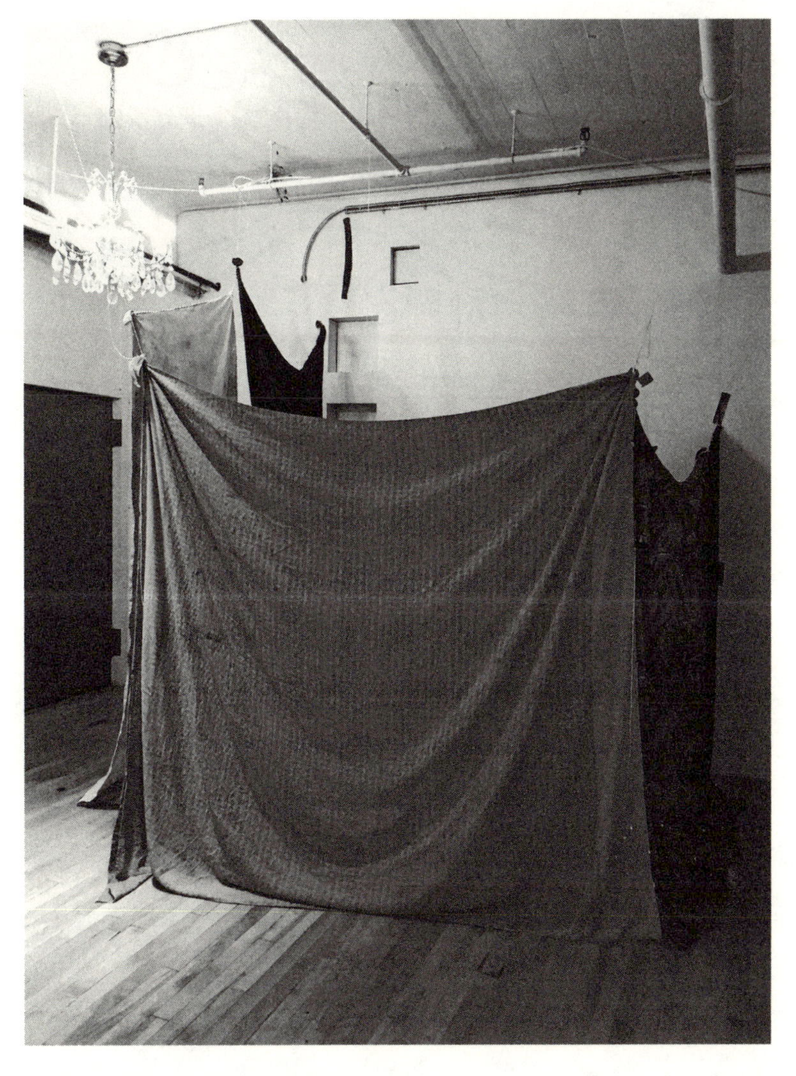

FIGURE 3.4. This tent, formed by hanging sheets, represented a "safer space" at a feminist hacker event. Montréal, Canada, August 2016. Photograph by the author.

to welcoming people who may feel marginalized in various ways, especially within wider technical cultures.

Separate spaces within hackerspace culture have formed in a number of cities, often as a result of forking existing communities when disagreements about the values and culture of a particular space could not be reconciled. As such, discussions about values and governance are salient at the outset, as referenced in the quote opening this chapter. One such example can be

FIGURE 3.5. Design for DIY flogger made from bicycle tube, 2008. Courtesy of Daniella Echeverria.

found in Philadelphia, PA, when in 2011 people left a hackerspace ("Hack-Make," a pseudonym) over fallout that occurred after disagreement over whether the space should host an event for hacking sex toys. Copy from the proposed event read, "If you like hacking and you like sex toys, then this is the event for you. From DIY floggers to vibrators that vibrate in Morse code, the possibilities are endless" (figure 3.5). Proposed by two women, one of whom was a board member of the space, the event was opposed by other members on several grounds, including that it might tarnish the public image of the space, which they claimed was especially important as the group was in the process of pursuing 501(c)3 status (charitable nonprofit organization). One member, Liam, wrote to the hackerspace's email list, "At the risk of coming off as a prude, I really don't like this idea. I don't think that it's appropriate to have this as an open house [event]. . . . This isn't the sort of public face I was hoping all our new visitors . . . would be seeing."[61]

Quickly, though, the conversation escalated to a full-bore argument over the values of hacking, which in part turned on diversity and inclusion. Another man wrote to the list:

> Expecting everyone to come into this group with the same goals, and the same plan to accomplish those goals, is the exact opposite of what this [space] is all about. . . . As we reach out, . . . diversity will come. The fact that we all seem similar now is because we are developing the 'core.' It's no coincidence [that current members] were looking for [this community, and] we found what we were looking for [at HackMake] and we all happen to be that 'type.' Call it primary requirements; geek, engineery, and employed. As we expand our outreach, so too will our membership's diversity expand, or not.[62]

On the one hand, the hackerspace member allows for variability among members' goals and even regards it as a high good. Different people wanting different things for the space should not present a problem. At the same time, he notes that "we all seem similar now" and says that this homogeneous constituency comprises the core membership of HackMake (along the lines of geek, engineer, and employed—which marks class, and implies gender). This is tacit acknowledgement that there are meaningful differences between the majority of the current members and "other" people, including those proposing the event. He assumes that greater variability in membership will follow in the future, primarily through reaching out to other types of folks (here unnamed).

Clara, the board member who had hoped to schedule the sex toy hacking workshop replied, disagreeing respectfully but vehemently. She wrote:

> There's evidence that the opposite is true, that once your core is established, diversifying is a losing battle. . . . I am not confident that HackMake would survive that process if it took it on in earnest in, say, a year. It takes a difficult analysis of what makes a place welcoming and who might

61. Email,—to [HackMake list], February 5, 2011.
62. Email,—to [HackMake list], February 5, 2011.

> be shying away without you ever knowing it. Without internal pressure to diversify, there are very, very few organizations that do it successfully even with targeted recruitment, which we're not doing. That's up to and including super left social justice movements that are consciously and actively anti-racist, anti-sexist, etc. HackMake is none of those things.[63]

The board member did not want diversity to be an afterthought, and was candid in her assessment that even with attention devoted to these matters, they would present an uphill battle. Notably, the sex toy hacking event was not proposed as a diversity event. But it became a flash point as it dredged up differing viewpoints about whose goals for HackMake should be given priority, and how central diversity should be to the mission of the hackerspace. One person who said he was ambivalent about the event itself framed the issue as one of not shying away from controversy: "I personally want to push boundaries of technology and society, make things cooler, and do things that make people think. I do not want to walk away from science, discussion or technology just because it's controversial."[64] Both these concerns—diversity and not shying away from controversial content—were salient and nearly inseparable in the debate as it unfolded. They drew out varied strands of thinking about the values within the space and how the community wished to proceed. This commenter also professed his commitment to "science and technology"—the hacker article of faith in technology as progress.

In the end, Clara's disgruntlement mounted to such a degree that she backed away from not only the event, but HackMake itself. She approached another hackerspace in Philadelphia and hosted the event there. These words, written before her decision to leave, show her struggling with the fact that her workshop suggestion turned out to be so contentious:

> So my concern here is that IT'S A HACKERSPACE. Initiative shouldn't be punished, particularly initiative that shakes up old patterns. [HackMake] is really stratifying into hardware tinkering as the core interest, and white males as the demographic.
>
> I'm frustrated in the extreme right now. . . . And I definitely don't want to flounce off b/c I can't [host] a particular event, but you can see how this is a culmination of concerns that have been building for months.[65]

63. Email,—to [HackMake list], February 8, 2011.
64. Email,—to [HackMake list], February 4, 2011.
65. Email,—to—, February 3, 2011.

Her statement that "initiative shouldn't be punished, particularly initiative that shakes up old patterns" is at the heart of the matter. Many people conceive of hacking and making as expressions of agency and even rewriting social order. It is therefore to be expected that, for some, the infusion of new kinds of people and new kinds of projects would be instantiations of this very agency. To be opposed by one's own community, in ways that seemed to uphold a social order of "white males" as the primary agents of technology, was experienced as a stinging failure and a moment of exclusion. Rather than expend further energies on "dragging people into caring" (in the words of Joe, the HOPE discussant, above), Clara opted to leave and regroup elsewhere.

And she was not alone. At least two other interlocutors in the debate who were active in membership and leadership roles in both spaces also exited HackMake in the fallout of this controversy. The new space, which welcomed the sex toy hacking event, has over time through the efforts of these people and others evolved into one that identifies as a feminist hackerspace with an inclusive mission. It hosts electronics, sewing, crafts, arts, and other activities for adults and for children. Its members believe that they are addressing "felt shortcomings of hacktivism" and creating a stronger hacking culture.

The details of this "great toy hacking kerfuffle of 2011" (as it was colorfully characterized by Clara later) are idiosyncratic as opposed to generalizable. A fallout over sex toy hacking is by no means the most common reason for forking hackerspaces. To an outsider, this episode borders on comic, possibly absurd. But it presents themes that are commonly at stake when members of contemporary voluntaristic technology communities wrestle with their values and raisons d'être. There is not infrequently dissonance over whether members see their activities as linked to "social justice movements that are self-consciously anti-racist and anti-sexist," or responding to "the core need [of] a group of folks who wanted to have large tools that were too expensive or inconvenient to own individually,"[66] as another person posted during the kerfuffle. This book argues that even when social justice goals are foregrounded, the link between those goals and hacking and making is often more emergent than self-evident. But in the context of the present chapter, the conflict illustrates that diversity was a core concern for some, who believed it could not be separated from the goals of hacking, and felt they needed to work within a community that unquestioningly shared this belief.

66. Email,—to [HackMake list], February 8, 2011.

This vignette also illustrates that much more than material goods are getting produced in open-technology communities. While eventually some electronic microcontrollers were attached to sex toys, this episode had other products. We also see the production of social relations, and a new infrastructure of care to guide relations. Not only did this episode increase investment in articulating the values of inclusive community at the second space, but members of HackMake also told me this incident caused introspection. Liam, who had opposed the workshop and stayed active at HackMake, told me he regarded some of the events of 2011 as "particularly embarrassing" and said that since 2011, he has been struggling to call out and prevent sexual harassment in the space.[67]

Also in 2011, an initiative formed to support women in open technology and culture. Active until 2015, its main activities included advocating for policies—including codes of conduct and photography policies—to prevent harassment at conferences, running "Ally Skills" workshops, and running unconferences. I attended two of the seven unconferences that it hosted: one in Washington, DC, in 2011, and another in San Francisco, CA, in 2012.

The unconferences were unstructured events that allowed attendees to self-organize into sessions on strategies that could promote inclusion in various open-technology spaces and projects. They were notable for the degree to which organizers focused on safety and privacy of attendees. This was an obvious reaction to the perceived climate for women in technology, as suggested above by the person who wrote, "I've been attending cons [conferences] of various types since I was thirteen, and I have never, not once, been to a con where I wasn't harassed." One measure the organizers took was to make the unconferences open only to women* (people who identified as women in some way that was significant to them), though this evolved over time. The Washington, DC, conference was also open to men, although the proportion of women was much higher. By the next year, the organizers had decided to limit the conferences to women* only, and to run Ally Workshops for men as separate events.

Organizers of these events felt strongly about privacy and safety in part because of the abysmal record of ugly incidents in various venues in open technology. The initiative had in fact been founded in part as a response to the experience of a young woman who reported that she had been sexually assaulted at an open-source conference in 2010. There are too many incidents to recount, but it is worth retracing the contours of one prominent instance of

67. Personal email, July 27, 2016.

controversy and harassment that occurred during the course of my research on this book. In 2013, at the PyCon conference in Santa Clara, CA, controversy erupted when a woman in attendance reported (via Twitter[68]) to the conference organizers that she was overhearing two other (men) attendees making what she thought were inappropriate jokes about "forking" (copying code, as mentioned above) and "dongles" (hardware connectors). Both of these terms can be made to have obvious, if puerile, innuendos. In deeming these comments inappropriate she took her cue in part from the conference's code of conduct, which explicitly forbade "sexist, racist, and other exclusionary jokes . . . excessive swearing and offensive jokes."[69] She included in her tweet a picture of the people making the jokes, presumably to aid conference staff in identifying the men. The tweet appeared in the stream of tweets generated at and about the conference; it was not a private message, and was visible to anyone following the conference hashtags.

Over the next few days, this incident metastasized and traveled beyond the conference setting.[70] One of the people implicated in making the jokes was fired by his company; though it is unclear what exactly occurred, and the confidentiality of personnel decisions makes it difficult to determine the degree to which the PyCon incident was a factor, the employer released a statement indicating the employee's conference behavior was implicated in the firing.[71] The fired employee soon surfaced on *Hacker News*, posting (under a pseudonym) an apology to the woman who had reported him, as well as giving "his side of the story."[72] This sparked much comment and debate, some of it inflammatory. Much more troublingly, unknown actors next launched a distributed denial of service (DDoS) attack on the software company for which the woman worked, bringing down its servers and crippling its client services. The woman herself was fired next, presumably because her company felt its operations were compromised by keeping her on staff. She was also *doxxed*. The term, which comes from *documents* ("docs"), refers to having private information like one's home address, personal phone number, and other personal information released. Naturally, doxxing is experienced as very threatening when it is directed against an individual. Uglier still, her doxxing led to implied and actual death threats, including Photoshopped images of her face on a bound and mutilated body.

68. Twitter is a short-message social media platform.

69. PyCon 2013.

70. See, for example, Marwick 2013.

71. PlayHaven 2013.

72. Mr-hank 2013.

The woman is African American, and the violent and threatening imagery and text directed at her was racialized as well as gendered. The total effect of the release of this woman's personal information along with being targeted with deeply violent, misogynist, and racist imagery must have been terrifying to experience. Tragically, it was not singular. Sarah Banet-Weiser and Kate Miltner refer to such phenomena as *networked misogyny*; they write, "We are in . . . an era that is marked by alarming amounts of vitriol and violence directed toward women in online spaces. These forms of violence are not only about gender, but are also often racist, with women of color as particular targets."[73]

I attended PyCon in 2011 and 2012 for fieldwork on this project, but did not attend the year of "Donglegate." As noted in the introduction of this book, I had identified the Python community as one where sincere energies were being devoted to diversity issues, and I had found my experiences at PyCon to be fairly congenial. I was thus somewhat surprised when social media and some of the electronic mailing lists to which I subscribed positively exploded in March of 2013, as the incident was beginning to unfold. Matters only escalated, as described above. It seemed rather ironic that the attack on this woman grew out of an incident in the Python community, where there was clear and concerted effort among many, including the organizers of PyCon itself, to be sensitive to these issues. However, this calls attention to the fact that awareness of these issues in open-source communities is no guarantee that the problems are under control. It also unclear, of course, how many of the people who piled on to harass and threaten this woman were active in Python versus being vigilantes from vaster reaches of the internet.

I recount the events of Donglegate here to illustrate why organizers attempting to support women in open technology have been sensitive to privacy and safety issues. Not all incidents are equally chilling. But neither are all of them as well known or as popularly exposed; it can be assumed that less-documented incidents have occurred, and this was certainly the attitude taken by the organizers of the unconference. One of the measures they took to make the unconferences safer spaces (in a literal sense) was to have a photography policy at the unconference sites that allowed attendees to indicate their level of comfort with being photographed. Photography was of particular importance because it was likely that participants would

73. Banet-Weiser and Miltner (2016: 171). See also See Lisa Nakamura's discussion of "glitch racism" (2013) and Beard 2014.

take the opportunity to photograph the unconference events for their own purposes including personal documentation, social media, and journalism; it was especially likely that in a gathering that celebrated agency in open technology and culture, making and sharing of media would occur. That said, organizers of the unconference were concerned about the use of photography to harass people, including the potential to aid people outside the conference in stalking or harassing attendees. For example, a photo on social media, which could have been posted innocently, might give away the identity and approximate real-time whereabouts of an attendee who did not want this information revealed. Ironically, it is possible that a more specific photography policy at PyCon would have prevented Donglegate from being set off.[74]

The organizers thus instituted a clear policy around photography, in which attendees' nametags hung from colored lanyards. Red meant "photos never ok, please don't ask"; yellow meant "ask first"; and green meant "photos always ok" (figure 3.6). All unconference participants could choose a lanyard that expressed their preference regarding photography by other attendees (figure 3.7).

This policy, about which organizers were adamant, caused friction when Google offered to host an opening reception for the San Francisco unconference in their downtown office (near the unconference site). Google's default for events was to ask everyone in attendance to sign releases consenting to photography, and it is obvious why an event for women in open technology might seem like a publicity opportunity for Google (figure 3.8).

Up until the very moment of the reception, unconference organizers were frantically trying to square things with Google, to make sure that photography would not occur unless people had opted in, or better still, to not photograph the event at all because the lanyard system had not been implemented at that site and photography preferences were not clear. According to organizers, multiple attendees had balked over Google's desire to photograph them and revealed their hesitance to organizers, whose commitment to making the conference a safer space ultimately trumped Google's public relations goals. This was somewhat awkward for organizers, as they accepted Google's provision of space, food, and drink for attendees while rejecting the usual quid pro quo of a public relations opportunity for their host.

74. Rianka Singh (2018) addresses differential vulnerabilities for people who are multiply marginalized participating in platform feminism, centering experiences of women who are brown, black, queer, trans, and/or disabled; see also Daniels 2015.

FIGURE 3.6. Red, yellow, and green lanyards, unconference, San Francisco, 2013. Photo by Flore Allemandou, CC BY-SA.

FIGURE 3.7. Unconference attendee sporting green lanyard that indicates consent to being photographed. San Francisco, 2013. Photo by Jenna Saint Martin, CC BY-SA.

FIGURE 3.8. The author's nametag from the Google-hosted reception. Note the female (Venus) symbol (in pink) in the Google logo. San Francisco, CA, 2013. Author photo.

Separate Spaces: The Backchannel

Related to concerns about locating and creating hospitable environments, online and off, diversity advocates devote huge energies to communicating about the culture of open-technology projects. Many conversations on smaller communities' mailing lists, social media, IRC, and the like are devoted to more and less open or public discussion of the reputations of various projects or individuals. They feed into, but are also distinct from, broader discussions. The conversations can take the form of warnings or of affirmations that a project is welcoming and that people have had good experiences there. These practices of documentation and backchannel communication are unsurprising in the context of a self-organizing ethos, and they are quite common in the online and offline spaces where smaller communities congregate to consider diversity issues. This communication itself composes an infrastructural layer that reflexively evaluates how open-technology projects are constituted. It potentially feeds into governance norms and structures within such projects, as examples below will illustrate. And it also has the capacity to steer new participants into projects or spaces, or steer them away. There is a nearly infinite amount of territory for these smaller communities to evaluate, and I can sketch only a small, fairly mundane handful of examples of their communicative practices.

On a mailing list including many unconference attendees, one woman reached out for advice, saying that she hoped "to understand and contribute to an actual system that actually does something,"[75] and wondering if others had recommendations of open-source projects that were open to new contributors. Another woman on the list, Anika, immediately suggested a couple of options, and then circled back to the list many months later presenting a fuller list of ideas that had come her way or that she had learned of subsequently. When she named the projects, she wrote,

> My criteria for recommending something here: it's open source, it's actually running somewhere and giving value, multiple people work on it, I believe the leaders are friendly and they help new coders contribute, and I know of no abusive behavior in that community. . . . Please feel free to reply all to say yay or boo to any of these, or to recommend other hospitable projects that are past 1.0 and that are actually running and usable![76]

Two issues stand out here that are relevant to the present discussion. First, though it is important that the projects be "running" and "giving value," both querent and respondent are fairly agnostic about what sorts of goals the projects might support. The assumption is that the main goal is to pursue the pleasure and opportunity of coding and learning in an environment that is ideally quite supportive, and certainly not hostile. What is being coded is almost secondary.

The second point is how Anika—more than the original poster—guides the inquiry toward not just a quick answer to the question posed, but to a collated list, available to list subscribers long after the first question had been answered to a satisfactory degree. As noted above, the impulse toward documentation is a strong one in these communities. This collation also indicates the respondent's conversance in what media scholar Carrie Rentschler calls "a politics of communicability" commonly seen in feminist technocultural networks.[77] Anika's response illustrates collective communicability in action: a network mobilized to praise and recommend projects with welcoming communities and to alert members of projects whose culture

75. Email ,—to [Women in Open Tech List] list, January 2, 2015.

76. Email,—to [Women in Open Tech List] list, June 30, 2016.

77. Rentschler 2014: 68.

is more questionable. (This particular discussion was nearly all of positive experiences, it should be noted.)

Documentation and collation serve multiple purposes here. First, and most apparently, they answer a question and maximize the replies to the original poster by offering a list of several projects. More subtly, though, the respondent's revisiting of the question across time gives the exchange new life, perhaps inviting people for whom the initial email thread had not been of use or interest at the moment it began to engage with the suggested projects, at or subsequent to the moment of recirculation. Leaving lists on display or archiving them for people to engage with at their own speed adds an important temporal element. Lastly, this collectivized communicability in a backchannel context serves to affirm, through ritual communication, the existence of the subgroup in the wider milieu of open technology. In other words, these forms of address within and directed toward the smaller community demonstrate its unity by voicing nominally common concerns. This is especially important for a community that is dispersed in space, discovering and communicating with each other via online communication: a concern they share (in the manner of Kelty's recursive public).

Many other examples of collectivized communication that encompass these three elements can be identified in the realm of diversity advocacy in open technology. The Geek Feminism wiki is a popular site and exemplar. Founded circa 2008, it is devoted to providing a place for feminist techies to come together. It joins the project of feminism with the culture and aesthetics of geekdom. Wiki pages address, for example, fan fiction, "recreational medievalism," and cosplay (dressing up as a character from a story, particularly anime).[78] The "About" page of the Geek Feminism blog says: "Things that are on-topic for this blog: 1. geeky discussion about feminism; 2. feminist discussion about geekdom; 3. geek feminist discussion of other things."[79] The wiki assumes conversance in topics such as "privilege, sexism, and misogyny" and recommends visitors to the site unfamiliar with such concepts start by reading before contributing to discussion.[80] It is home to a regularly updated "timeline of sexist incidents" in geek communities, which features a list and explication of many reported incidents of harassment or bad behavior in tech communities, including industry and open source. It includes a description of Donglegate in 2013. Like the emails about

78. Geekfeminism.org n.d.

79. Geekfeminism.org, n.d.

80. See Reagle 2016.

"hospitable projects past 1.0," the wiki serves to inform and is a resource across time. However, even if transmission of information is the manifest reason for these wikis and emails, for the diversity advocacy subgroup it is also apparent that there is a salient ritual element to these communicative practices that finds itself constituted in these exchanges of attention.[81] Similarly, in 2014, a new mailing list was begun to share information about incidents in hackerspaces where people had been asked to leave and not return. Its initiators wrote, "In the interests of collaborating with other hackerspaces on safe space issues myself and [Name] have set up an inter-hackerspace mailing list to share information about folks that are not welcome in spaces as a potential warning to others."[82] Though digitally distributed and at core united in technological affinity or geek identity, these practices resemble those of consciousness raising established by feminists in the 1970s, which also included self-defense and even crisis counseling. These communicative acts also perform a folkloric function; they maintain collective memory and group identity through the circulation of accounts.[83]

Instances of these sorts of communication abound and are too numerous to recount. In part, they underscore just how much terrain there is to evaluate and monitor, including conferences, hackerspaces, IRC channels, and mailing lists. In May 2016, a man in Poland wrote repeatedly to a global Python electronic mailing list to report and discuss troubling behavior on a Polish IRC channel for Python. He reported on several occasions that numerous people in the channel were making racial, homophobic, and misogynistic slurs (in Polish, the language of the channel; he provided the original language and translation when he wrote about these incidents). He himself replied to those in the IRC channel that there was no place for that behavior on the channel, but his efforts to intervene made little difference and even led to him feeling threatened. As he wrote (in English) to the list, "There was warning that if I'll continue what I'm doing—'backstabbing [his interlocutor in the channel]' [which probably meant reporting channel problems there and to #python-ops, a channel for operators, people with power to enforce behavior], it will end very badly for both sides. I don't know

81. Schiller 2007.

82. Email,—to [Hackerspace discuss list], November 24, 2014. Unrelatedly, someone made a general comment to another mailing list that expressed that she understood the need for such measures but hated relying on them: "In a community if I have to use backchannels to just survive and work for free, then something is seriously wrong and that is an unequal relationship" (Email,—to [Women in Open Tech List], July 31, 2015).

83. Jones 1991.

how to interpret 'ending very badly', but it sounds like a threat."[84] Other list members replied to encourage the poster who was reporting the bad behavior, and wondered whether the Python Software Foundation (PSF) itself might intervene to clarify rules for IRC channels. One person wrote an eloquent email reminding the poster why his efforts were needed, even though he at times felt like giving up:

> Someone may ask why making [this channel] a friendly Python developers channel, where different people are respected and feel welcome is so important? Why not just simply ignore the channel and let it rot in the toxic atmosphere? Well, the problem is that someone new to Python in Poland sooner or later usually finds [the official Polish Python] website, . . . and there he/she finds a link to [this] IRC channel. So from time to time different new people appear on the channel, usually young ones, who watch the channel and start to believe that derogatory comments, racists jokes or laughing [at] women developers is something natural and accepted in Python groups. How many of the new ones, from those oppressed groups will soon after that leave, believing and spreading information that Python developers are complete jerks? That's why I think what you are doing is very important, even if it seems . . . like talking to the [channel's] operators and abusers is like talking to a wall.[85]

For this poster, the only way the Python community could grow was to be mindful of how new participants would experience the community. Thus, while attempting to change the culture on this particular IRC channel—one lone online space, important mainly to Polish-speaking Python users—might seem tedious or even futile, the poster saw great import in these efforts, for Python specifically, and, it can be inferred, for open source more widely.

The fellow in Poland trying to address uncivil behavior in the channel saw the situation partially resolved when the channel decided to adopt the PSF code of conduct the next month. To adopt a CoC was not a prerequisite for having a channel devoted to discussing Python, which anyone could do, but it sanctioned the particular channel and allowed it to keep a channel name that gave it "primary status" for the topic at hand, Python in Poland. In other words, unofficial channels could certainly exist, but they were not endorsed by the PSF. As usual, concerns existed that the CoC was only as good as its enforcement, and the poster in particular was worried that

84. Email,—to [diversity list], June 15, 2016.

85. Email,—to [diversity list], June 13, 2016.

some of the channel's operators (people with greater administrative power, including the power to ban people) were not as militant about enforcement as he would like. That said, the original poster reported that the behavior was at least somewhat improved, and as far as he could tell the CoC alone was responsible: "On June 26th PSF CoC was officially adopted on [this] channel anyway. After that the level of aggression on the channel significantly decreased. . . ."[86] Once again, we see the significance of backchannel communication, both to offer instrumental advice and support to the poster in Poland, but also to affirm the value of inclusive behavior in the Python community to readers of the electronic mailing list.

"No More Rock Stars": The Backchannel Goes Public

In late 2014, a backchannel electronic mailing list discussion began about a public statement issued by the Tor Project. Tor is a prominent free software project that enables anonymous communication by relaying a user's internet traffic through a network in such a way that makes the traffic's source impossible to trace. Funded by private foundations and the US government, it has been at the center of controversy for its support of internet privacy. In the statement, Tor expressed "solidarity against online harassment."[87] Members of the electronic mailing list expressed a variety of reactions. One person wrote, "I don't understand what it's supposed to mean, really. It's basically just a bunch of people publicly expressing they don't support online harassment."[88] Another person agreed, writing, "They make it look like they are taking action somehow, which they aren't—and that's what I find weird/don't like about it."[89]

Others weighed in to note that this statement might signify a wider change in discourse. Note that this discussion was occurring only months after "Gamergate," which had begun in 2014. Though a discussion of Gamergate is beyond the scope of the present chapter, it was a remarkable, conspicuous, and vile instance of online and offline harassment of women in gaming. It is likely that Gamergate, among other events, was on the minds of those crafting Tor's statement. One person replied to those who didn't see why they should care about the Tor statement:

86. Email,—to [diversity list], July 12, 2016.
87. Tor Project, 2014.
88. Email,—to [Women in Open Tech List], December 15, 2014.
89. Email,—to [Women in Open Tech List], December 15, 2014.

> The context to me is one where that community often positions "anti-harassment" [on] one side, and "anti-censorship" or free speech on the opposite side. The Tor folks are contradicting that, and saying they can both fight *for free speech*, and *stand against harassment and misogyny*. It is a big change for an organization in this field to even admit that women, people of color, trans people, and others who are marginalized undergo more, or different kinds, of harassment than straight white men.
>
> So while this is not suggesting any practical solution, it's a good step towards taking a stance, or at least an attempt to shift their culture.[90]

Another person agreed. She wrote,

> I also see the [Tor] statement as the beginning of a potential normative cultural shift among certain hackers since they clearly state at the organizational level that online harassment is unacceptable. Such statement is in my opinion crucial in trying to continue to craft and influence social and cultural norms within the hacker and tech communities.[91]

It is apparent from these responses that diversity advocates viewed the Tor statement with some degree of ambivalence or even skepticism, noting that it did not translate into especially meaningful action. At the same time, they also provisionally accepted the statement as a potential harbinger of a wider cultural shift. The poster who writes that to see "anti-harassment" and "free speech" on the same rhetorical side within the culture of advocates for computer freedom is "a big change" is indeed correct that this was a relatively new development, discursively. The other poster is also correct in noting that Tor issuing this statement *as an organization* represented a step toward institutionalizing this norm. Both posters expressed provisional hope that Tor's statement indicated that dearly held beliefs of diversity advocates—which had not heretofore been mainstream beliefs in open source—might take hold in the wider culture.

That said, many on the list expressed serious reservations over whether Tor was a model for projecting an inclusive culture or curtailing harassment. Among diversity advocates, it was something of an open secret that Tor had internal problems for which this public statement could in no way compensate. As one person pointedly but euphemistically described the situation to the mailing list, "There are multiple stories that one of their (male) developers has

90. Email,—to [Women in Open Tech List], December 15, 2014. Emphasis added.
91. Email,—to [Women in Open Tech List], December 15, 2014.

some personal problems with the idea of consent."[92] It was widely believed that a prominent Tor developer had abused his position of power within the company by crossing lines in his personal behavior, including sexual misconduct. This colored reception of Tor's stand against online harassment, with some people believing it was at best ineffectual and at worst hypocritical.

I relay this backstory because it presaged events that occurred a year and a half later. In summer 2016, a series of public accusations made against the prominent Tor developer cited in this 2014 email discussion rocked hacker, open source, and infosec (information security, digital rights, and privacy) communities. Numerous people came forward to allege that he had mistreated them, with some of the allegations rising to the level of sexual assault. The developer resigned from Tor. This episode sparked an intense reaction, much of which is beyond the scope of this book.[93]

Some of this informs a discussion of infrastructures of care. What is striking is how diversity advocates seized upon this episode to urge greater reflection in open source about governance and leadership. Though the allegations in the Tor case were extreme, many people felt that this shed light on patterns in open-technology projects that were common. In particular, they seized upon the alleged abuser's "rock star" status within Tor and related computer privacy and internet freedom communities. Many felt that in order to strengthen open-technology projects, star status—for anyone—would need to be curtailed. A long blog post entitled "No More Rock Stars" expounded on this tenet and included several specific guidelines for voluntaristic open-technology groups to consider. These included:

Have explicit rules for conduct and enforce them for everyone
Insist on building a "deep bench" of talent at every level of your organization
Flatten the organizational hierarchy as much as possible
Distribute the "keys to the kingdom" as much as possible[94]

As previously noted, Tor case was extreme: it involved multiple public allegations of sexual misconduct.[95] But these governance rules should be under-

92. Email,—to [Women in Open Tech List], December 15, 2014.

93. See Marechal 2018 and West 2018 for discussions of how crypto communities formulated an analysis of intersectionality meant to inform both technical best practices and discussions of governance. See also Poster 2018 on how cybersecurity can benefit from women's expertise.

94. Honeywell 2016.

95. This community had unique and complex reasons for choosing to accuse the developer publicly but not involve law enforcement. Because of activist commitments, including anarchism,

stood as aiming to curtail abuses of power long before they could rise to such an egregious level. They sought to prevent any particular individual from becoming indispensable to a project, and to disallow any individual from having enough control over any particular domain to consolidate power. The author of the post acknowledged the "tyranny of structurelessness" critique and did not advocate shelving hierarchy entirely. She advocated for having an accountable, transparent hierarchy, which was as horizontal and distributed as possible, instead of having unofficial leaders who governed projects through charisma, reputational capital, or technical prowess.

These issues were tied to diversity in the minds of people familiar with the situation and seeking to change open-technology cultures. One of the people who had come forward as a victim wrote in a blog post, "We would likely have had more diverse contributors to Tor, if we had dealt with [Name] sooner, since, for years, many people have been warned about [Name] through a whisper network and [dissuaded] from becoming involved."[96] This person posits that more women would have been likely to become involved in Tor if it were not an open secret that Tor had these internal problems. (We can see evidence, in the 2014 email exchange, of Tor having a reputational problem centering on this developer's alleged behavior.)

Technically oriented cultures tend to exalt technical expertise and are therefore potentially set up to experience power differentials along the lines of this expertise. In particular, people who embody expertise tend to be celebrated for it, and this is especially true when their social identities match the implicit notion of who is a technical expert (often white men, like the person accused of misconduct in Tor). Another person who had also come forward as a victim in the Tor community said that she thought a culture of valorizing technology itself, and especially granting certain people star status through their technical prowess and willingness to claim credit for technical accomplishments, was part of the problem. She wrote, "The 'shut up and code' mentality creates a hostility to basic human empathy—talking about feelings, offering support, engendering individual and collective growth—which could have helped with accountability and prevention."[97]

---

many were distrustful of police and criminal justice pursued through the state. They were also concerned that activists working in computer privacy were often targeted by states through the criminal justice system and did not wish to have their accounts used as fodder for state persecution of a fellow activist; they aimed for restorative justice instead.

96. Lovecruft 2016.

97. Macrina 2016.

Along related lines, Dorothy said in an interview that open-technology culture would, in her opinion, benefit from rethinking leadership. She felt that in open-source projects in particular, "the same people have a chokehold over all the decisions."[98] As an African American woman, she felt that the decision to essentially grant "leader for life" status to project leaders stifled participation and in particular made it harder for people less likely to be considered as leaders to gain recognition for their contributions. This essentially created a vicious cycle where more minoritized people had a tough time getting a toehold and being recognized, which could breed mistrust or disillusionment, causing these people to accept diminished roles or quit the projects, in turn rendering projects less diverse and consolidating power within the existing structure. She said, "how refreshing it would be to be a member of a project where they rotate leadership." However, she was pessimistic that many project leaders would be willing to give up their current status or roles, characterizing this attitude as, "'Oh my god, if we let them [new people, and new kinds of people] have leadership roles, they are going to change the organization to how they want it.' Well, aren't *you* running the organization how you want to run it?"[99] Though her comments are general and address slightly different aspects of diversity in open source than those made by the people commenting on the fallout in Tor, many of her recommendations run in a similar direction. Like the "No More Rock Stars" author above, Dorothy wondered what could be gained by changing governance patterns such that positions of power rotated, and that no players were in a position to make decisions in perpetuity. This points to reimagining sociality as a key intervention by diversity advocates, who seek to change the terms of association within open-technology projects and cultures.

Conclusions: Progress, Ambiguity, and Maintenance

Diversity advocacy in open-technology culture is a vast project. Throughout this chapter, I have illustrated multiple types of infrastructural interventions undertaken by diversity advocates at multiple sites. The terrain that open-technology social infrastructure spans is nearly infinite, spanning IRC, electronic mailing lists, social media, conferences, and hackerspaces

98. Interview, Dorothy, August 2016. Shaw and Hill (2014) argue that peer production organizational modes can function as "laboratories of oligarchy," wherein a small group of early members consolidate and exercise a monopoly of power over others, not the laboratories of democracy often attributed to peer production.

99. Interview, August 14, 2016.

and makerspaces, yet most interventions are painstaking and highly local. This illustrates the problem of scale: even if advocates' care is pitched at a grandiose level, the hacks and patches that they can reasonably undertake are necessarily smaller in reach. As the chapter has chronicled, some salient efforts included raising awareness, advocating for codes of conduct, starting separate spaces, engaging in backchannel communication, and proposing changes to democratize project governance. My aim is not to evaluate these interventions for their efficacy. Rather, I draw attention to them as significant in representing measured attempts by a subgroup within open-technology culture to change the terms of association within that culture. In these interventions, both forking and agitation occur: diversity advocates tend to answer the question "Do we change the culture of this group or start our own space?" with the reply "both/and."

To be sure, these are delicate and complex undertakings. It must be stated that just because diversity advocates strove to foster safer spaces, this goal was not necessarily accomplished and would be difficult to quantify in any event. All of the online discussions I refer to in this chapter included at various points expressions of concern that unknown or explicitly untrustworthy people had access to the discussions, even those that were supposed to be relatively private, among known and trusted people. There was also a high degree of awareness of the near-impossibility of keeping electronic group communications truly private; people were often warned not to be overly trusting of the security of electronic mailing lists, for example.

In addition, concerns over trust and inclusivity could be influenced by insider versus outsider status that included, for example, racial overtones. In an interview, Dorothy disclosed to me that she believed that some "white feminist" diversity advocates had on more than one occasion sided with white men and against her; she said she "wondered who the white women feel solidarity with, that they would view me [as an African American woman] as an extremist."[100] By contrast, she wondered if some other diversity advocates viewed all men as "unsafe" or potentially untrustworthy. She said she herself sometimes had an easier time finding allies in men than in white women. These issues illustrate the challenges of autonomously creating and maintaining safer spaces, including potential pitfalls in building coalitions and common cause across categories of social difference. Just as there is an active, vocal diversity advocacy community questioning the

100. Interview, Dorothy, August 2016. See Gray 2016 for discussion of similar dynamics in gaming.

wider culture, there are also agitational voices within that subgroup, issuing critique and providing a reminder that not only technical cultures but diversity advocates' interventions have the potential to be experienced unevenly. Interventions aimed at the wider culture still carried the burden of the social identities and roles of the participants, even if some of that burden was what people hoped to shift.

Some of the issues with which diversity advocates wrestled were undoubtedly made more complicated by the fact that they were intervening into *technical* communities. Historically, the "priesthood"[101] of technical expertise has been occupied by people with greater social power, often men, whites, and other elites. As one discussant had provocatively claimed in a panel discussion at a hacker conference, in spite of commitments to horizontality, hackerspaces have often tended to be elitist spaces. And *technical* expertise has been held in higher regard than any other, to the extent that other aspects of open-source cultivation should be subordinated to it ("shut up and code"). Diversity advocates mounted challenges along two lines. First, they insisted that more effort should go toward recognition and valuation of people in technical communities who did not fit preconceived notions of the social identity of "technical expert"; the contributions of members of various minoritized social groups needed to be made visible and then normalized. Second, they argued for more recognition of people who contributed to projects in ways other than coding—to focus too much on technical expertise was to normalize a hierarchy that devalued a bevy of other roles that kept projects and spaces running. These two critiques were interrelated, of course. But because of their relationship to technical expertise, technical cultures potentially pose unique challenges for those who seek to democratize them.

In addition, open-technology cultures also possess unique characteristics that made diversity advocacy difficult. As one person wrote in an email inviting participation in a workshop for women, trans, and queer people in Amsterdam, "hackers & activist communities are proud of being anti-authoritarian and using non-mainstream approaches to legality & rules."[102] While it is certainly not the case that hacker and open-technology cultures are devoid of cultural norms, it is true that rhetorical commitments to freedom, for example, historically have lent resistance to diversity advocates' insistence on norms of civility and "caring." This is why it was potentially

101. Marvin 1988.

102. Email—to [Europe List], June 23, 2016.

important that the Tor project brought "free speech" and "anti-harassment" (in advocates' words) closer together in a platform of internet freedom.

To bring these goals together was no small challenge. The connection had to be actively forged and then maintained. G., the founder of a respected open-source project that Python diversity advocates repeatedly claimed had fostered an inclusive environment, said in an interview that he understood his own role as engaging "the privileged white boys" in conversations to underscore that "we're not changing the rules here, we're just asking for a moment's extra consideration."[103] G. said while he did this over and over, the need for it constantly surprised him. He told me that he was most amazed and troubled not by the "sociopaths" in the hacker community, who are well known for flagrant antisocial behavior online (much of it virulently sexist, racist, and homophobic), but by what he called the "second wave"[104] of programmers' attitudes. He said these were people who were "otherwise reasonable," but who nonetheless, in his words, were "hemming and hawing, justifying the foul-mouthed misogyny, who are [saying], 'well, it's true that more [of you] women should get into programming, but on the other hand [I reserve the right to call you] a stupid cunt.' "[105] To his mind, this attitude needed to be shifted. He felt he, as a white man and respected figure in open source, was uniquely positioned to help with this.

But in spite of G.'s claim, diversity advocates were indeed trying to change the rules. His rhetoric probably represented a compromise, or a feint; to rework sociality within open-technology cultures was risky and challenging. In a coda to the situation of the Polish Python IRC channel, the person who had drawn attention to inappropriate behavior in the channel wrote once again to the global Python email list, a few months after the code of conduct had been instituted. To his dismay, jokes about pedophilia, gibes at the CoC itself, and swastikas[106] had all been posted in the channel

103. Interview, July 3, 2012.

104. This is a reference to a continuum of attitudes toward women in open source, not to "second wave" feminism, though his use of this term probably indicates conversance with that label as well.

105. Interview, July 3, 2012.

106. Discussion ensued on the electronic mailing list as to whether this was "actual Nazism" or mere "trolling" of the CoC. Observers seemed to think it was probably the latter but agreed that the distinction was irrelevant as it far surpassed boundaries set by the CoC. This illustrates how often extremely offensive and hateful speech can be trotted out under the banner of free speech (online and otherwise); it also illustrates how the most convenient vocabulary for "offensive" speech is often insults tailored for marginalized groups.

within the last few days.[107] He was concerned that the operator tasked with enforcing the CoC was either not up for the job, or perhaps privately disagreed with it being applied in the channel. An administrator not active in the channel but who possessed the power to kick people out temporarily or ban them permanently wrote back trying to get a handle on the nuances of the situation. That person wrote that one danger of having an outside, higher-up administrator take action is that such a move

> still only further reinforces the idea that the people who want a CoC are "them" and the people who don't are "us." If it turns out that that's not just how a few jerks who [resent the CoC] perceive it, but actually an accurate representation of [reality amongst the users of the channel], we [advocates for diversity and civility] are in a lot deeper, because now we need to go find a bunch of [new] operators for a channel that is already not doing too well.[108]

In other words, the administrator wondered if intervention from outside the Polish Python channel's own administrators might backfire and succeed only in driving a wedge between the Polish channel's users and the rest of the Python community that had tenuously agreed upon the value of a CoC for the whole community. His worry was that if it seemed like a user ban had been imposed from outside, this might solidify unity among the channel users who thought the CoC was undesirable.

This presents a new problem, not raised earlier. While many had expressed worries over whether CoCs were useless if not enforced, here there were diversity advocates monitoring the use of the channel, who were willing to take enforcement action. However, enforcing the CoC from outside risked admitting the channel was ungovernable from within, all the while not changing the minds of people who found it acceptable to post swastikas. Infrastructural intervention had resulted in an ambiguous stalemate. This was undoubtedly not the last word. But it plainly illustrates the pitfalls of autonomous governance within open-technology projects: the need to agree upon and maintain rules and standards, and the relative chaos that ensues when diversity advocates intervene into wider technical cultures. The larger story here is that, though diversity advocates' ambitions are to affect sociality at a large scale, sometimes the absolute most they can do is hold the line in a single space or project.

107. Email,—to [diversity list], September 4, 2016.
108. Email,—to [diversity list], September 4, 2016.

The underarticulated politics of open source, useful in some circumstances, leave its practitioners ill-equipped to best manage current calls for diversity. If their political ethos is one of classical liberalism, it should not matter if some people want to depart to hack a new project or space: there is ample room for the pursuit of many ideas. On the other hand, there is a communitarian impulse that indicates that a greater good—and stronger movement—will emerge from everyone hacking together. In other words, to fork is not a solution if the goal is to make the community work better for all participants. There is a striking tension between hacking as meritocratic undertaking and the need to think about seemingly extraneous topics such as *who's in our group?* or *who is winning our arguments*? Technical enthusiasts who are devoted to technology in part because it seems to offer a rarified domain that is separate from messy problems such as social identity are suffering a paradox in their worldview, which is difficult and complicated to resolve. *Are these groups producing goods or are they producing communities, social relations? What rules and infrastructures produce the best communities?* What is going on here is not just a tweak of norms but a form of hacking that is congruent with how geeks solve technical problems: they are metaphorically adding code to their communities and debating how the new artifacts function.[109] A problem they have not resolved is who gets to decide what the best design is. Diversity advocacy magnifies the tension between individual pursuit and communitarian ethos, rather than diminishing it.

109. Kelty 2008. Thanks to Lucas Graves for discussion.

4

Crafting and Critique

ARTIFACTUAL AND SYMBOLIC OUTPUTS OF DIVERSITY ADVOCACY

On a warm, muggy afternoon late in August in 2016, approximately ten people assembled in a large, windowless, dimmed room inside a loft art space in Montréal. They were participants in a five-day "feminist hacking convergence." For the afternoon, they trained their energies toward choreographing and enacting a *cryptodance*. As one might deduce from the context and name, this brought together cryptography and dance. Less apparent is what this could possibly be.

The cryptodance was an experimental event that conjoined arts practice with pedagogy about the principles of cryptography in computing. Peculiar though it may seem, there is precedent for rendering scientific principles in the idiom of dance.[1] The initiators of the cryptodance described it as "a performative event to familiarize ourselves with different modes of encryption. Whilst collectively embodying issues of security, privacy, safety and surveillance, we converge a technopolitical agency for souvereignty [*sic*] and a desire for affinities with the body/machine—living organisms/algoritms

1. Whether or not its initiators were aware of it, the cryptodance calls to mind "Protein Synthesis: An Epic on a Cellular Level," a "happening" produced by the chemistry department at Stanford University in 1971, featuring dance, music, and an introduction by future Nobel laureate Paul Berg. (For a description of this spectacle, and a 2009 "hemoglobin dance," see Myers 2015: 219–21.) Thanks to Will Schofield and Cyrus Mody for discussion.

FIGURE 4.1. Setting up electronics for aural component of cryptodance experimental event. The person in the foreground is wearing a device that purportedly converts electrical impulses on her scalp into sound. Montréal, Canada, 2016. Author photo.

[*sic*]."[2] Encryption has recently gained a high profile within activism, policy, and intelligence circles as they debate privacy and security in electronic communication, including chat applications, SMS (short message service), and email.[3]

Astras, the Belgian collectivity leading the cryptodance, had spent several hours over preceding days preparing electronic sound components to accompany the dance (figure 4.1). On the day of the event, they readied participants by reading aloud from a text they had prepared about the history of encryption technologies. This narrative included a description of Alan Turing's research on encryption and persecution for his homosexuality in the era of World War II, as well as an explication of steganography. They then enrolled me, as a native English speaker comfortable projecting my voice, to

2. Cryptodance n.d.

3. See West 2018.

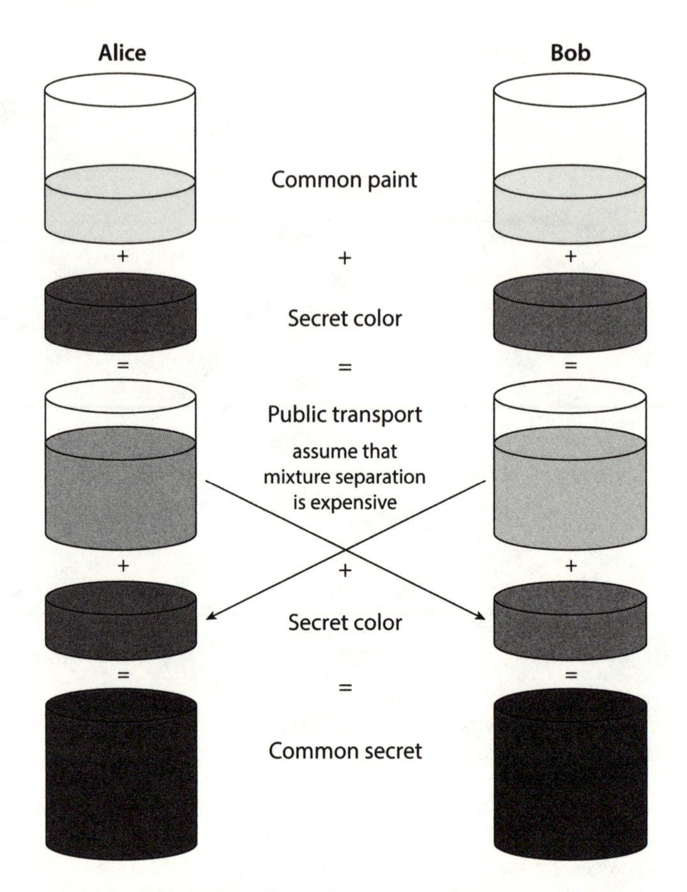

FIGURE 4.2. Model for Diffie-Hellman key exchange, 1976. See DuPont and Cattapan 2017.

read some of the text to the other participants (for many of whom English was a second language). They improvised the instructions for me, asking me to read not only text from a page but then asking me to describe images on the page that the rest of the audience could not see, which appeared to be photographs of people participating in a somewhat similar event, but were essentially indecipherable to me and assumedly more so to my audience. Next, they played a 2016 video clip in which Canadian prime minister Justin Trudeau opined about quantum computing. This somewhat inscrutable narrative built until they had concluded the historical introduction, with a diagram of the Diffie-Hellman model of encryption projecting onto a wall (figure 4.2).

Next, the cryptodance participants, who had up until this point been sitting on the floor, were instructed to rise and follow a series of instructions to move around the room. The lights stayed low, and the room began to thrum with electronic sounds. Many of the sound-producing components were interactive, such as a headband that converted electrical impulses from the scalp into sounds, which supposedly varied as the wearer's thought or movement patterns changed (figure 4.1), and circuit boards with soldered leads hanging off them that one could place on one's body or other conductive material to create resistance in a circuit, the sound of which would be amplified and played through a speaker. One of the instruments was a theremin in need of repair, which was plugged in but unresponsive to nearby hand movements. (Later that weekend, in a workshop on soldering, people got the theremin's lower frequencies working, though it was not fully restored.) The digital projection changed from the Diffie-Hellman diagram to a feed from a camera placed at floor level, showing us moving in the room, but mainly feet and lower bodies. The camera, we were assured, was not recording.

Everyone in the room was enrolled into cryptodancing. One of the first instructions was for people to partner up and one person to lead while the other followed. The leaders were instructed to move around the room, across the floor space, and also shift planes of height: crouching, standing. The leader was supposed to guide the follower, whose eyes were closed, through space by holding her hand. Somehow I wound up in a cluster, Astras leading myself and Aure, a researcher/dancer from Québec. In recovery from orthopedic problems in my left hip, I closed my eyes and hesitantly shuffled around the room, feeling the floor with my feet for cables I might trip over, minding my balance, while trying to follow Astras's lead in terms of when I should be close to the ground or more erect. I could sense that Aure was more at ease with this exercise; as a dancer she moved confidently across the room. Astras ably balanced our different levels of comfort, and no one tripped, fell, or radically broke formation. Next, people opened their eyes and returned to maneuvering the room solo, following instructions to keep their gazes on objects in the room (some of which might be stationary but probably moving if the object was a participant). Then, we were instructed to add interaction: incorporate an element where our action would feed back into the room. The squawks and squalls of the electronics took on increased urgency, and people moved deliberately into the frame of the camera projecting our movements. Someone extended a hand toward the theremin and an ethereal squeal pierced the room. This seemed a response to her hand, but was a coincidence, as the theremin was nonoperational.

Having suitably warmed up to the rather strange, kinetic, and experimental nature of the event, the cryptodance proper ensued. The Diffie-Hellman

FIGURE 4.3. People developing the cryptodance, with their moves projected on the wall. Montréal, Canada, 2016. Photograph by the author.

key exchange model went back onto the wall, and we broke into two groups (representing "Alice" and "Bob") and began to interpret Diffie-Hellman using our bodies (figure 4.2). Each group was tasked with figuring out an embodied means of representing the elements in the model. People improvised, and then memorized, a series of dance movements that comprised the model, assigning iterative moves to each component. As suggested by the model, the steps blended together. Then Alice and Bob swapped half of the dancers (as "public transport"), who had to teach their group's elements to the other group (figure 4.3). The resulting synthesis, collectively embodied by all participants, was the "common secret." The dancers watched themselves projected on the wall to make sure everyone had learned the correct common moves.

Once the cryptodancers had gone through the routine a few times, the dance broke up and people relocated to sit outside at tables in the street below, breathe in some fresh air, and reflect on how the dance had gone. People local to Montréal planned to reprise the event with new participants a few months later in their home city; our cryptodance was both a rehearsal and its own event. The Belgian collectivity would lead versions of the event back in Europe as well.

The cryptodance provides a unique point of entry into considering the material outcomes—in the form of artifacts and products—of the practices of diversity advocates.[4] Some are more tangible than others. In many cases, like the cryptodance, products are emergent, fragile, or speculative. Though this chapter takes seriously the material aspects of the practices and artifacts under consideration, it argues that the products' primary significance arises from their symbolic dimensions. This includes what they telegraph about their origins: how they are produced, and by whom. Rather than standing alone as material products or practices, they represent a range of critical stances—sometimes prefigurative or underarticulated—formulated in reaction to mainstream open-technology communities. This is very similar to the "recursive public" discussed by Christopher Kelty, whose members participate in "reorienting knowledge and power" around technology, and through their practices create a public.[5] He writes that "free software geeks argue *about* technology, but they also argue *through* it."[6] Similarly, diversity advocates in open-technology projects critique mainstream open-technology communities through acts of production. Their products and artifacts become a means of argumentation and address among themselves. These artifacts affirm identities and values within their own subgroup, which is similar to what political theorist Nancy Fraser calls a *subaltern counterpublic*.[7] They may also signify this subgroup's existence to a wider open-technology public. Material practices and artifacts signify belief in a cultural order that poses a challenge to the prevailing one.

4. Thanks to Judy Wajcman for putting this simple question, "What gets built?" to me. I have continued to think about how to answer, which is less simple.

5. Kelty 2008: 6–7.

6. Kelty 2008: 29, emphasis in original.

7. Fraser 1990. A *counterpublic* is a discourse community marked by its subordinate status to, and some degree of exclusion by, a wider, dominant public (Warner 2002).

Though the cryptodance was an ephemeral product, largely improvised and evanescent, it has significance as an instantiation of diversity advocates' beliefs about how to imagine and embody alternative relationships between people and technologies.[8] The general description of the convergence read, "[This] event aims at addressing the lack of women, queer, trans and diversity in technological fields in general and hacking more specifically. But even more so, it aims at creating a community that critically assesses the hegemonic narratives around technologies."[9] The convergence's conveners proffered a critical orientation to technological fields (including their composition) as the convergence's raison d'être. The organizers list the attributes of cryptodance: "communicating what cryptography is; emanicapation [*sic*] to develop trust; open form of discussion; support each other, solidarity," with one of its goals "technical familiarizartion [*sic*] as a political stance." This statement brings together technical knowledge (understanding cryptography; technical know-how) and an orientation that foregrounds trust, support, and solidarity among participants. Thus, the products and artifacts produced in the course of diversity advocacy cannot be separated from the political orientations and emphasis on criticality formulated by diversity advocates. They do not stand alone; rather, they are inextricably tied to the critiques being promulgated by diversity advocates.

As this chapter will show, the products and practices of diversity advocacy are worth sustained consideration for what they reveal about the borders of care for diversity advocates. Material objects, the most obvious product of hacking and technological production, are produced. But not all are instrumentally useful, and many ephemeral objects are produced in these spaces as well. What unites these products and practices for diversity advocates? It is evident that while material stuff matters, it is not all that matters. This is a crucial point for understanding the goals and outcomes of diversity advocates' efforts.

Making practices in the open-technology spaces where diversity advocacy is occurring have commonalities with what Matt Ratto has called "critical making." This is "a mode of materially productive engagement that is intended to bridge the gap between creative physical and conceptual exploration," which Ratto presents as a pedagogical strategy.[10] Maker practices in these spaces and projects do not only embody ordinary hacker/maker

8. Duncombe 2007.

9. Email,—to [Feminist Hacking List], June 1, 2016.

10. Ratto 2011: 252.

agency in opening up technologies. They simultaneously represent expressions of critical agency surrounding the issue of diversity, as they play with forms and practices of making that affirm, through material engagement, the presence of the subaltern counterpublic. In other words, this counterpublic is called into being in part through the material engagement its members undertake. As anthropologist Barbara Myerhoff has written, "One of the most persistent but elusive ways that people make sense of themselves is to show themselves to themselves . . . by telling themselves stories. . . . More than merely self-recognition, self-definition is made possible by means of such showings."[11] This chapter argues that acts of material and artifactual production in diversity-focused open-technology projects often have a symbolic dimension that is at least as important as the material outcome. They are telling stories with artifacts as opposed to words, because these artifacts show the subaltern counterpublic to itself. As in other instances of critical making, whether or not the products of diversity advocacy making are instrumentally useful, they promote reflection and conceptual exploration. Here, they also foster in-group solidarity. These dimensions of making in diversity advocacy are *at least* as important as material, artifact-based outcomes.

Having used the cryptodance as a point of entrance into the products and artifacts that diversity advocates create, I sketch a few more examples below. All should be taken as representative in a broad sense, though each is situated in its own particular context.

Hacking Meets Feminine Craft

A prominent feature of many of the products created by advocates of diversity in open technology is how they conjoin traditionally feminine craft practices with electronic components. Textiles that incorporate the LilyPad Arduino kit are a prime example.

Developed by MIT designer Leah Buechley and commercially introduced in 2007, LilyPad is an Arduino variant geared toward use in e-textile contexts. Arduinos, as discussed above, are microcontrollers that can be programmed in a variety of ways. They are sold commercially but released as open hardware, meaning the design files are available for people to modify the design and share their modifications. They are usually programmed using open-source software.

11. Quoted in Orr 1990: 187.

Buechley observed that soon after the introduction of LilyPad, sales figures, in the tens of thousands, indicated that approximately 35 percent of the buyers were women, and 57 percent men. This was a striking difference from the purchasing statistics for the standard Arduino (from the same retailer), which indicated that men accounted for 78 percent of sales and women only 9 percent.[12] Over 90 percent of customers were in Europe or North America.[13] As these numbers show, the LilyPad design was significantly more popular with women, who represented just over one-third of its purchasers, than the regular Arduino, for which women were less than a tenth of the consumer base.

The design of the LilyPad lent it to use in textile projects, as the kit included conductive thread that could be used to configure the circuit and to anchor it to a fabric base. One use suggested by the designer was to use the LilyPad to program LEDs on a jacket for use as bicycle turn signals (figure 4.4).

In a design such as this one, the microcontroller is sewn onto the garment using conductive thread. The LED lights and their switches are also sewn to the garment. Because the thread is conductive, it permits a working circuit to be mounted on the garment, which allows for various possibilities incorporating electronics into wearable items. Notably, these electronics can be programmed and attached to their bases without wires or soldering.

The significance of the LilyPad exceeds the application of this particular circuitboard or any specific project that incorporates it. In combining hobbyist electronics and sewing, it locates hobbyist electronics as being within the purview of feminine craft. This is important because electronics has a long legacy of being a masculine domestic hobby, as we saw in chapter 2. To reimagine hobbyist electronics within the feminine sphere is a way of expanding the identity of hacking and making (and of hackers and makers).

This should not be overstated, however. Computing has been associated with elements of artisanal practice for decades.[14] In spite of their subsequent obscurity in popular imagination, the programmers of the earliest electronic computers were women.[15] And women's labor has been integral to the global production of electronics. For example, in the context of RCA's hiring practices for consumer electronics production in Mexican *maquiladoras*, "Management's standard explanation for its preference for young female workers

12. Buechley and Hill 2010: 4.
13. Buechley and Hill 2010: 3.
14. Ensmenger 2010a; Coleman 2012a.
15. Light 1999; Abbate 2012; Hicks 2017.

FIGURE 4.4. The LilyPad Arduino (top center of back) controlling LED turn signal, 2008. Courtesy of Leah Buechley.

typically rested on the idea that women's mental and physical characteristics made them peculiarly suited to the intricacies of electrical assembly work."[16] Managers have often drawn parallels between needlework and the dexterity and attention to detail that electronics assembly required, citing "nimble fingers."[17] Computer historian Paul Ceruzzi gives a more detailed description of these fabrication processes, writing, "On an assembly line (possibly located in Asia and staffed by women to save labor costs), a person (or a machine) would 'stuff' chips into holes on one side of the [circuit] board.

16. Jefferson Cowie quoted in Nakamura 2014: 921. Of course, it is not lost on managers in global firms that young women workers of color can be paid less.

17. Nakamura 2014: 920.

She would then place the board in a chamber, where a wave of molten solder would slide across the pins protruding through to the other side, attaching them securely to the printed connections."[18] As these accounts illustrate, this work is often offshore and invisible to consumers, which explains in part how computing could be so long associated with white masculinity in the North American and European markets where consumer electronics flourished. By the 1980s, electronics assembly "was not just women's work, but women of color's work."[19] One person at the Montréal convergence said she had learned to solder as a child, helping her mother assemble circuit boards for her job.[20]

Nonetheless, in the realm of hobbyist craft, *soft circuit* designs sparked the imaginations of people interested in "hacking hacking,"[21] opening up electronics practices and artifacts to nontraditional electronics hobbyists such as women. *Soft circuit* is another term for circuitry conducive to being integrated with textiles.[22] (Needless to say, *soft* is also a signifier of femininity.[23]) At the 2012 unconference for women in open technology held in Washington, DC, one of the unconference tracks was on soft circuits. People discussed possible applications of soft circuit textile designs. One attendee, who was pregnant, kidded that she should make a fetal monitor for the HOPE conference (Hackers on Planet Earth, discussed in chapter 3), in which sensors in her clothing could give her feedback on "what sessions the baby likes."[24] Other people joked about a "proximity sensor dress" that would blink more aggressively as someone got closer to the wearer. This could mean either warning off someone whose approach was unwanted, or heightening the experience in a flirtation or romantic encounter. Another person said it could be combined with an eye-tracking sensor in order to convey the message "quit staring at my boobs."[25] These comments were not sincere attempts at designing material artifacts (though echoes of the sex-toy-hacking workshop can be heard in the idea for a garment sensor that tracks a lover's approach). Instead, they should be read as in-group signification, celebrating mutual affinity for feminine craft and customizable

18. Ceruzzi 2003: 193; Qiu 2016; see also Irani et al. 2010.
19. Nakamura 2014: 920. See also Haraway 1991a for early and foundational critique.
20. Fieldnotes, August 20, 2016, Montréal, Canada.
21. Rosner and Fox 2016; S.S.L. Nagbot 2016.
22. Peppler et al. 2014.
23. Edwards 1990.
24. Fieldnotes, July 10, 2012, Washington, DC.
25. Fieldnotes, July 10, 2012, Washington, DC.

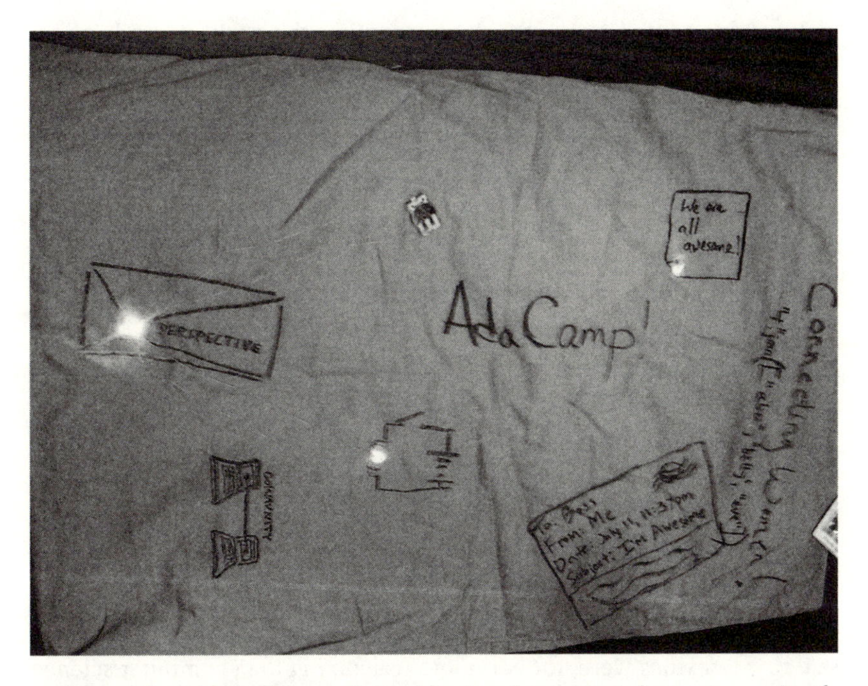

FIGURE 4.5. The finished pillowcase. Text and drawings show connected computers captioned "community," and a note reading "To: Boss, From: Me, Subject: I'm Awesome." Note the battery pack and three illuminated LEDs. July 2012, Washington, DC. Photograph by the author.

electronics. They also signified in-group membership through commiseration about common experiences for women in many technical spaces ("quit staring at my boobs").

In the session devoted to soft circuits, an attendee from Philadelphia had brought some supplies to quickly throw together something that would illustrate the principle of soft circuits. In the approximately hour-long session, attendees decorated a pillowcase with magic markers. They also stitched on a small battery unit and a few LEDs with conductive thread. The end product was a hastily assembled collective work that celebrated women in the workplace, community around computing, and the unconference (figure 4.5).

Though celebratory of soft circuits' potential, unconferencers also reflected on whether DIY electronics might contribute to reinforcing gender stereotypes in problematic ways. One person said she had long struggled with whether it was patronizing to try to bring women into electronics through sewing, but that she had ultimately concluded it was "not patronizing to meet people where they're at."[26] She also said that she found how

26. Fieldnotes, July 10, 2012, Washington, DC.

makers often touted Arduinos—"you can do anything with an Arduino"—very abstract. She said, "But what does this really mean? This comment is going to sound gendered, but I think a great intro would be 'hack your oven.' Find a device you already use, take out the proprietary crap that controls it, and control it yourself."[27] She said this approach could be taken to many household items, including the freezer, alarm clock, oven, and washing machine, newer models of which are all loaded with electronic controls. In practice, this would require not only a fair amount of expertise but a willingness to open up and break your appliances—not necessarily a welcoming prospect for a beginning hobbyist, but probably an enticing challenge for more experienced hackers and makers. Many people expressed enthusiasm for hacking automated watering systems and their household appliances. But one person said, "I need not beginner Arduino but pre-beginner Arduino,"[28] which again underscores an expertise barrier for true novices to electronics.

A young woman from Philadelphia who had recently graduated from college said that DIY craft could be an opportunity for hackerspaces. She said that e-textiles were, for her, more welcoming than computer science classes: "I took computer science classes but only got excited when I made my own biking jacket with turn signals"[29] (as shown in figure 4.4). Another person, also from Philadelphia, did not disagree but said that she thought that "a lot of hackerspaces are afraid of becoming craft spaces."[30] Though she did not spell out the gendered dimensions of this, they are clearly there. A woman from Montréal laid this out more plainly, saying, "A lot of the hackerspaces [in Montréal] are less than 20 percent women, and when new women arrive, all the guys tell the women to go talk to the new woman who came in."[31] Thus, while unconferencers did not dismiss bringing feminine craft practices into existing hackerspaces, integrating them and giving them equal footing with other practices and artifacts would probably be challenging and require delicate maneuvering—more effort than some of them wanted to make. It is unsurprising that feminist hackerspaces and other separate spaces for women have made traditionally feminine craft practices fairly central activities.

27. Fieldnotes, July 10, 2012, Washington, DC.
28. Fieldnotes, July 10, 2012, Washington, DC.
29. Fieldnotes, July 10, 2012, Washington, DC.
30. Fieldnotes, July 10, 2012, Washington, DC.
31. Fieldnotes, July 10, 2012, Washington, DC.

FIGURE 4.6. Artwork advertising a feminist hackerspace zine-making event, 2014. Courtesy of Hannah Schulman.

To drive this point home, I turn to a zine-making event held at a women*-only hackerspace in San Francisco in 2014 (figure 4.6). It may seem surprising that a hackerspace would routinely host a low-tech event like zine-making, but we should not view this as incongruous. As Alison Piepmeier has noted, "Zines created by girls and women . . . are sites where girls and women construct identities, communities, and explanatory narratives from the materials that comprise their cultural moment. . . . These documents . . . are [also] a site for the development of late-twentieth-century feminism."[32] Insofar as a central ambition of the hackerspace was providing a site for the cultivation of a feminist-identified subaltern counterpublic, the practice of zine-making there was fully consonant. This became more apparent in the discussion about zine-making that evening. The evening's activity was to make a zine that reflected "who comes here"[33]—in other words, that narrated the hackerspace's constituency. To accomplish this, people cut up magazines and pasted the images to 8.5 × 11-inch pieces of paper, which they also decorated with drawing and text (some written, some cut-out letters). Each page was to represent the person who made it, and they would be bound together to collectively portray the hackerspace.

32. Piepmeier 2009: 2–4.

33. Fieldnotes, June 22, 2014, San Francisco, CA.

The weekday evening workshop was attended by around fifteen people, many of whom had come straight from work. Most everyone appeared to be in their twenties or thirties, and while people did not uniformly present as white, a majority did. Nearly half sported a shock of vividly dyed hair, a common feature of people in hacker and tech scenes, alternative culture, and fan culture, as well as at intersections of those cultures. The event's hosts asked people to introduce themselves to one another. Introductions were to include one's name, one's preferred pronoun (several people said they preferred "she/they"[34]), and to tell the group "how you're feeling, if you saw something weird today, and your favorite animal."[35] Though this was in some ways a fairly generic introduction and icebreaker, and it was certainly meant light-heartedly, it struck me that no one had asked me to disclose to a group of people my favorite animal since I accidentally accompanied a friend to a church youth group's after-school meeting, at around age eleven. Between the favorite animal and the confession of how one was feeling, the zine-making activity was marked as one where a sort of atavistic girlhood or adolescence was on display and welcomed. Zines can be understood as public-facing diaries, also a feminine, confessional genre. This introductory exercise also exhibited hints of a more direct feminist sensibility: in the disclosures of weird events of the day, people not only gravitated toward random mishaps and amusing urban juxtapositions, but one or two instances of unwelcome masculine attentions (firsthand or observed), in keeping with feminist cultural practices of witnessing.[36] (This event occurred a couple of years before the cascading #MeToo moment began.)

The evening's participants also swapped ideas and attempted to recruit one another to create content for other zines they were working on outside of the evening's workshop. One person said she was seeking submissions for a zine about occult, "witchy," or magical things.[37] Another said she was taking submissions for "awful stuff said about your body," inspired by a recent date who had told her she was "very sturdy."[38] A number of zines

34. I have made a sincere effort to appropriately gender people in this book and use preferred pronouns. However, especially given that some people's preferences change over time, the lag times between initial drafting, revising, and publication, and that I do not know the full or proper names of all of the people who appear in this book (making following up with them challenging), it is possible that some people have been misgendered, for which I apologize.

35. Fieldnotes, June 22, 2014, San Francisco, CA.

36. Rentschler 2014.

37. Fieldnotes, June 22, 2014, San Francisco, CA.

38. Fieldnotes, June 22, 2014, San Francisco, CA.

were passed around and on display for perusal or inspiration. Members of the hackerspace had previously worked together, producing a zine about health told through the stories of vaginas in a family and one about menstrual periods. This latter zine included a "hipster period tracking app" that was a paper chart; it was meant to wink at the Quantified Self movement so popular in the Bay Area tech circles.[39] Notably, all of these proposed and extant examples show the circulation of knowledge about women's bodies, including less-credentialed forms of knowledge such as witchcraft. Whether or not the occult zine-writer's interest in magic should be taken at literal face value, witchcraft symbolizes a mode of resistance to patriarchy and the subjugation of women's bodies.[40] According to cultural historian Janice Radway, zines "function as a critique of the marking and individuation of girls' bodies as *visible*—that is, as bounded objects that can be controlled through others' surveillance. At the same time, they constitute an alternative strategy for making girls present *to* and *with* each other."[41] In feminist zine making, forms of knowledge like folk medicine can be filtered through the riot grrrl practice of zine-making, which is itself connected to long traditions of feminine papercraft and journaling. They are identity practices in addition to circulations of knowledge.

My final example of feminine craft cum hacking is a 2012 piece of artwork—*Compulsive Repurpose*—produced by a hackerspace founder and artist in Philadelphia (figure 4.7). The piece features a knit textile, rather resembling a homemade scarf, rendered mainly in Ethernet cables. The first cable is connected to a computer keyboard, and the artist leaves visible the craft and production process by including the chunky knitting needles in the piece.

Explaining her rationale for producing the piece, the artist wrote,

> I think I had recently learned about a project that aimed to get more women into hardware hacking, specifically the Lilypad Arduino and the work of Leah [Buechley]. She developed the Lilypad because many women are familiar with sewing and the materials involved, and figured it could get more women using the Arduino. . . . I thought for myself, as I struggled to find an entry point into hardware that was interesting, what if I combined a craft that I already like (knitting) with hardware. The Etherknit piece was the result. I think it did open up a lot of avenues

39. Lupton 2016; Nafus and Neff 2016.

40. Federici 2004.

41. Radway 2010: 224.

FIGURE 4.7. *Compulsive Repurpose*, knitting project using Ethernet cables, 2012. Courtesy of Georgia Guthrie.

> I was interested in exploring, like how to use the inner wires of the ethernet cable to add circuits with LEDs. . . . Other people responded well to the piece, and I would be interested in pushing it further sometime in the future.[42]

This quote shows the conscious fabrication of a link between electronics and craft practices such as sewing and knitting likely to be already in women's skill set; in the artist's own words, she used knitting to find an entry point into hardware. It also emphasizes that to produce speculative objects like *Compulsive Repurpose* is to communicate, "other people responded well to this piece." Sometimes the objects themselves can forge connections, create juxtapositions, and spark ideas in ways that augment or exceed other forms of communication about the same concepts. These objects not only introduce new materials and practices into hacking, they "hack hacking" by challenging its boundaries. That said, what is effected by these practices is far from clear.

42. Personal email, September 8, 2016.

Autonomy of Infrastructures

Another category of product that is worth scrutiny is *autonomous infrastructures.*[43] Once again, this can range from speculative to concrete things. Autonomy is a key component of both hacker and cyberfeminist beliefs about technology, which are complex but have roots in the Appropriate Technology movement of the 1970s and 80s, and in the older American notion of self-reliance.[44] It also echoes the small producer ethos of indie and punk music scenes, and DIY, as practiced—particularly in Europe—within squatter movements and autonomist political movements.

The *feminist server* is an artifact that exemplifies autonomous communications infrastructure. Owning and running their own servers has long been a priority for geeks of various stripes, including those whose political beliefs cause them to fear censorship or interference.[45] Control and access to online content is affected by what server is used, which becomes an important consideration for groups doing political organizing. For example, on the radical left, the group Riseup devotes server space to mirroring content for social justice organizations on at least three continents. It is also a badge of geek pride to keep servers running under challenging conditions, or in unusual environments; anthropologist Gabriella Coleman writes that it was "delectable and engaging" for Debian neophytes to learn project folklore such as a key server once being housed "under x's desk in his Michigan dorm room."[46]

But feminist servers take this geek pride and interest in political autonomy in a new direction, infusing feminist critique. My first introduction to feminist servers was in a conversation with Pala, a Belgian designer I met in New York City in 2014. Wanting to learn more afterwards, I looked them up online. I found a website discussing the topic, which contained an intriguing list of principles for design and operation of a feminist server:

> Is a situated technology. She has a sense of context and considers herself part of an ecology of practice

43. Fieldnotes, August 19, 2016, Montréal, Canada. "[Autonomous infrastructures] seem to share a desire to create the conditions for their autonomy in terms of their governance models, the values they embrace and the principles they promote," write the conveners of feminist hacking events (https://thf2016.noblogs.org/summary-of-thf-2016/).

44. See Pursell 1993; Turner 2006.

45. Writing of the prehistory of Indymedia, Todd Wolfson writes, "[People] recognized that with a server came a greater degree of independence to create their own Web presence and support other groups in taking full advantage of the Internet" (2014: 52).

46. Coleman 2012a: 52.

Is run for and by a community that cares enough for her in order to make her exist
Builds on the materiality of software, hardware and the bodies gathered around it
Opens herself to expose processes, tools, sources, habits, patterns
Does not strive for seamlessness. Talk of transparency too often signals that something is being made invisible (Division of labour—the not so fun stuff is made by people—that's a feminis[t] issue)
Avoids efficiency, ease-of-use, scaleability and immediacy because they can be traps
Knows that networking is actually an awkward, promiscuous and parasitic practice
Is autonomous in the sense that she decides for her own dependencies
Radically questions the conditions for serving and service; experiments with changing client-server relations where she can
Treats technology as part of a social reality
Wants networks to be mutable and read-write accessible
Does not confuse safety with security
Takes the risk of exposing her insecurity
Tries hard not to apologise when she sometimes is not available
This is not a closed 'definition'. Other principles can be added to this list of principles.[47]

It was not clear to me whether this list of principles described a real object.

In point of fact, multiple feminist server projects launched in the mid-2010s, in Europe and Latin America. One that began in Amsterdam and later migrated to Graz, Austria, was described more prosaically: "[This] server is run by women, using free software only. It acts as a place to learn system administration skills, host services and inspire others to do the same."[48] Feminist hardware hosts feminist content: for example, feminist servers in Europe host backup content for groups in Latin America that includes information on reproductive health—particularly abortion—as Latin American

47. Anarchaserver 2014.

48. Syster server, 2015. Its name, "syster server," references the same play on words as the Systers email list serv, begun in 1987 ("sister" plus "sys admin"). See also Metzlar 2009.

activists are fearful for the security of their content on domestic servers due to the perceived sensitivity of the information.[49]

That said, the principles listed above gesture toward a function for these servers that is also more speculative and highly symbolic, and that borrows heavily from feminist theory. They acknowledge frailty (risk exposing insecurity; try not to apologize when unavailable). The servers are autonomous, but not along the lines of a notion of autonomy that foregrounds heroic self-reliance ("She decides for her own dependencies"). Their technology is recentered around social relationships ("Treats technology as part of a social reality"; "Builds on the materiality of software, hardware, and the bodies gathered around it"; are run by a caring community). And the servers are understood as situated, located in a place, a community, and a context; this is a challenge to universalist narratives about technology.

Autonomous infrastructure is imagined in numerous ways. During the feminist hacking convergence in Montréal, two local activists presented to attendees their plans to launch a new collective oriented around computer training, with an emphasis on salvaging discarded machines, getting them running, and recirculating them within the local community. They related this to autonomy in multiple ways. Most importantly, by reclaiming discarded machines, the collective hoped to establish meaningful independence from the computer production chain, which brought with it dependence on a global supply chain including offshore labor and environmental practices they did not wish to support. Though they celebrated computers as essential tools, they did not want to uncritically adopt the associated practices of consumption. They were also opposed to sending computers out into a refuse stream that used labor and environmental practices they found troubling. In addition, they hoped to "disrupt the cycle of planned obsolescence."[50]

The activists said they were modeling their collective on other groups that existed in a few cities across North America. The (quite ambitious) scope for the project included taking in donated machines from the local community, refurbishing them, installing free software, allowing people who volunteered a set amount of time to take home a machine, and donating working machines to social justice groups. They hoped to create a "support system for people to know and use free software," holding open community events where "anyone can get their computer fixed and help running it as

49. Fieldnotes, August 20, 2016, Montréal, Canada.

50. Fieldnotes, August 20, 2016, Montréal, Canada.

long as it's running Linux."[51] They had spoken to people at recycling centers that took in donated electronics about getting their hands on unwanted machines, which were plentiful. They hoped for feedback from the feminist hacking convergence about refining their goals and taking next steps.

A couple of issues stand out here. One is the scale of activist ambition versus the reality on the ground. Though it was the case that Montréal, like other cities in the Global North, could supply activists with a steady stream of discarded consumer electronics like PCs, there were other resources that were harder to come by. One would be a stable space to host a repair shop and workshops. Astras (who was around forty, and thus had more years of hacking and activism under their belt than the Montréal activists, who appeared to be in their twenties) immediately suggested that the activists "factor in mobility [of the equipment]" and suggested that they assemble all the work stations on wheels, under the assumption that a grassroots technology project just getting off the ground might change venues multiple times before finding a stable home, if it ever did. Astras also suggested proactively splitting up the available spaces to suit needs, perhaps using one site for storage and another for teaching or workshops; they said this would make the collective "not so reliant on current infrastructural possibilities."[52]

Questions about space and sustainability recurred for diversity advocates seeking to carve out a hackerspace or repair shop of their own. Discussing an Amsterdam group devoted to women in hacking, begun in the early 2000s, a primary member recalled in 2009 how they borrowed spaces to meet, one in a kitchen of a squat, and another in the office of an open-source firm. She said, "Those spaces were free, so that gave us the opportunity to be somewhere. And for the rest we did all this on our own time as volunteers, so that's free, but obviously people need to have that free time to be able to do it."[53] She also discussed the first server the group used to host its content: "The one server was donated by the company that one of the [our collective members] was working at, and then when [we] had problems with that . . . we managed to find space [on another server, gratis], so the only thing we pay for is the DNS [domain name system registration]."[54] Both the face-to-face meeting space and the server space were borrowed or donated resources

51. Fieldnotes, August 20, 2016, Montréal, Canada. See McInerney 2009.

52. Fieldnotes, August 20, 2016, Montréal, Canada.

53. Metzlar 2009.

54. Metzlar 2009.

that were available because participants were situated at the intersection of paid IT work and hackerspace social worlds.

One of the most extreme examples was in New York City in 2015, where a few people were trying to get a feminist hackerspace off the ground. This effort ran for over a year as a monthly after-work meeting for people interested in becoming members of the aspiring space. That summer, meetings were held in the warehouse space leased by the Etsy corporation, the e-commerce site marketing handcrafted goods, which was located in the DUMBO neighborhood of Brooklyn, just across the river from lower Manhattan. Upon arriving, in the dwindling heat of a brutally hot day in late August, I realized I recognized the building from having attended artists' open studios events there in years past; the area had largely transitioned from a marginal arts neighborhood taking advantage of fallow warehouse space to one of luxury condominiums and corporate offices.

The Etsy space was provided under the aegis of an employee. At this meeting, a handful of women worked quietly on coding projects on their laptops, breaking to eat dinner when food they had ordered was delivered. They all appeared to be in their twenties and white. A couple of them apologetically announced to the group that they had forgotten to bring their laptop power cables, so they could only stay as long as their batteries held charge. One said she had meant to bring a knitted cowl to which she was affixing programmable LEDs, but had forgotten it when she ran out the door early that morning for a doctor appointment and then headed straight to work. I had never attended a less hands-on hacking or tinkering meetup, but the reason for this became clear in considering the space in which they were convening (figure 4.8). They couldn't leave their projects there, as there was no storage space allocated for them. There were some tools and workbenches belonging to Etsy, but the effort required to bring material objects from home or another workspace, then to one's paid work, and back again, limited projects to highly portable ones. The founder of the hackerspace, who was not the Etsy employee, said she could offer a conference room at her workplace, a software company, "but we can't hold soldering projects there—or we could, but we'd have to be quiet about it, and we can't burn the table, I'd be responsible."[55] Etsy's space, where one could solder and sew, was better, but only marginally. Use of Etsy's space was contingent on the employee's presence; if she had another obligation, or was ill, the rest of the group could not use the workspace after hours. She commented, "I can't

55. Fieldnotes, August 31, 2015, New York, NY.

FIGURE 4.8. Aspiring feminist hackerspace members gather in the warehouse offices of Etsy, Brooklyn, New York, August 2015. Note the antique sewing machine atop an antique card catalog, against the far wall. Photograph by the author.

let you in here without an Etsy employee present, so your exposure to risk is high since you have to go through me, even though Etsy is supportive of fostering marginalized people in tech."[56]

Months later, the Etsy employee's words had proved prescient, and the connection had proved too tenuous to continue. The meeting moved to the corporate offices of Meetup—a software company whose online social networking platform facilitates in-person meetups for people in shared communities of interest—located in the trendy SoHo neighborhood of Manhattan. Again, after-hours access was contingent on the presence and commitment of a Meetup employee who was trying to launch the hackerspace. Not a year later, the nascent hackerspace's monthly gathering had relocated once again to another space, back in DUMBO—this time to the offices of a software company whose mission was to support independent and leftist media companies. As before, the spaces were fragile in their contingency and left little room for leaving tools or projects on site until the next meetup. In New York

56. Fieldnotes, August 31, 2015, New York, NY.

City, the high cost of real estate meant that the hackers were wholly reliant on sponsored spaces, and corporations were the entities by far most likely to control desirable warehouse or office space. At the time of this writing, while more than sixty people have expressed interest in joining the hackerspace and the in-person meetings have continued, the hackerspace does not have a permanent home. Members might have found more permanent quarters had they reached into outer boroughs instead of meeting in lower Manhattan and adjacent waterfront Brooklyn, but this would have presented logistical challenges to meeting after work; the aspiring hackerspace members could easily stop off in these neighborhoods on their commutes.

I include this discussion of spaces because products are affected by their producers' access to work spaces. In spaces where one cannot solder or leave behind a project in progress, material products are less likely than less tangible ones.[57] Of course, New York City in the early twenty-first century is an especially daunting place to carve out an autonomous space. Nonetheless, pursuit of autonomy of infrastructure often means carving out impermanent spaces that are highly contingent—often borrowed from corporations or other sorts of organizations—and being prepared to move. These enterprises are thus fragile, and the difficulty of scaling up projects that revolve around tools and artifacts that are challenging to move is clear. In multiple ways, such enterprises must situate themselves in regard to employment relations, and they are often far from autonomous.

One notable expression of autonomy as a value can be found in an Oakland, CA, makerspace led by people of color.[58] The makerspace itself was located in a largely African American, Chicanx, and Latinx neighborhood, in an old storefront space, a couple of doors down from a nonprofit bike collective, also run by people of color, with a particular emphasis on youth empowerment. The makerspace organizers said that in the (famously stratospheric) real estate environment of the San Francisco Bay area, the landlord gave them a break on rent in order to support their mission, which afforded them more stability than the NYC group. When I visited the space, I

57. As Bratich notes, craft is a form of affective production. He thoughtfully links craft to an etymological root meaning "power, strength, and might," in the sense of capacity or ability rather than domination or force (2010: 311).

58. I mainly use "people of color" where actors did; I do not wish to conflate all non-white-identified people, whose experiences with both tech cultures and race and racism can vary widely. Particularly in black communities, there has been recent concern about how grouping "people of color" together potentially contributes to black erasure, "a gesture of solidarity and respect [that has evolved into] a cover for avoiding the complexities of race" (Hampton 2018).

observed a vestige of an earlier meeting, in the form of a list on a whiteboard. Titled "To Learn for Apocalypse," the list read:

Running
Plant identification
First aid
Fighting
Herbal medicine
Starlight navigation
DIY clothes
Surgery with everyday objects
Archery
Hunting
Compost
Finding shelter
Water[59]

Whether or not this list should be taken at face value as an expression of the most prominent goals of the makerspace, it is nonetheless illuminating. It reflects the countercultural or New Communalist heritage of digital utopianism, especially prominent in the Bay Area.[60] Given the list's setting, it also echoes community self-sufficiency measures taken by the Black Panthers, whose impulse toward survival had a different cast than that of many relatively socially and economically privileged people who moved "back to the land" in the 1970s.[61] Autonomous infrastructure here dispenses with computing entirely, imagining survival in a postelectronic, broken, and forbidding landscape.

At the Montréal convergence, one person observed, "Autonomy of infrastructure in terms of software is not so bad—[but] in hardware we are far from it."[62] It is worth unpacking this quote. As discussed above, activists envisioned repairing and rebuilding hardware in line with their vision for autonomy from production and postuse chains to which they objected. They also struggled to find stable spaces in which they could conduct this work. Software, however, is worth considering as its own distinct case. As

59. Fieldnotes, July 24, 2014, Oakland, CA.
60. Turner 2006.
61. Nelson 2013; Turner 2006.
62. Fieldnotes, August 20, 2016, Montréal, Canada.

Christopher Kelty has argued, free software "allows values and principles to be turned into material objects, things that can be manipulated, reconfigured, tested and torqued. . . . [It] seems (or perhaps seemed) to have a peculiar power to leave this materialism radically open to change."[63] Kelty rightly points to the fact that software, especially software with its code left open, is always potentially ripe for reconfiguration toward different material or political ends. This property is a singular reason why code has been understood as a unique object for intervention, especially for those insistent on software freedom (which should, of course, not be confused with a wider political agenda of emancipation, let alone a progressive or radical politics).

The Montréal activist who made the comment was referring to FLOSS in general, including the famous Linux operating system. But we can see how her comment could apply in the case of Dreamwidth, a software project that reflects both an open-source ethos and a commitment by its founders to changing open source in terms of whose contributions were valued and allowed to constitute the project. Founded in 2008, Dreamwidth forked the code of LiveJournal, an online platform created in 1999, which is something of a hybrid between a blogging platform and social media: users blog, journal, and post other forms of content, and form communities of interest in which they follow one another's content.

Dreamwidth retains many of LiveJournal's features, especially the basic foundation of the site: users publishing and sharing content. The founders of Dreamwidth (both had been LiveJournal employees) were interested in having a site with more privacy and author control over content. They removed the advertising features that LiveJournal relied on for support and instead used subscriber revenue to fund the site; since its inception, Dreamwidth has been supported financially by users, not by advertisers.[64] It is maintained by a small paid staff as well as a much larger pool of volunteer developers. It is for-profit, but not especially growth-oriented; its leaders are content to cover costs, ensure reliability, and turn a small profit instead of focusing on ever-growing expansion. Two-thirds of the company's profits were

63. Kelty 2013.

64. A page about Dreamwidth's "guiding principles" reads: "[Advertising] changes the site's focus from 'pleasing the userbase' to 'pleasing the advertisers'. We believe that our users are our customers, not unpaid content-generators who exist only to provide content for others to advertise on. We are committed to remaining advertising-free for as long as the site exists." Dreamwidth, "Guiding Principles," Accessed October 13, 2016. https://www.dreamwidth.org/legal/principles.

earmarked for developments, with half of that amount put toward developments chosen by the Dreamwidth user and developer community.[65]

Dreamwidth is an exemplar of open technology in multiple ways. First, and most obviously, it runs on open-source software. The code it was built upon was open source, and the changes and innovations its developers have introduced are also open source: "We strongly believe in Open Source and the Creative Commons. By freely releasing all of our code and documentation, we hope to create a vibrant and thriving community that will constantly improve the product we offer."[66] The founders' commitment to Creative Commons (a proposed alternative to copyright law, which allows people to license their creations in ways that explicitly permit derivative works) in addition to open source reflects the values of the project and its user and developer community. Dreamwidth describes itself as "not fandom-specific, but fandom-friendly."[67] Its founders and users are emphatically committed to supporting fan fiction and fan artwork. The connection between open source and fan culture is well established; it is not uncommon for people to learn programming in order to support their fan enthusiasms.[68] And as noted above, the Dreamwidth user and developer community is given an explicit stake in determining how the project will proceed in terms of building new features and making changes.

Lastly, Dreamwidth is "open" in its explicit and catholic commitment to diversity, which founders embraced from the outset. In 2009, it adopted a diversity statement, which underscored its commitment to serving users (and rejecting advertising, as it plainly stated it would not privilege certain demographics over others). It read, in part:

> We welcome people of any gender identity or expression, race, ethnicity, size, nationality, sexual orientation, ability level,[69] neurotype, religion, elder status, family structure, culture, subculture, political opinion,

65. Dreamwidth, "Business FAQs," https://dw-biz.dreamwidth.org/332.html. Accessed October 13, 2016.

66. Dreamwidth, "Open Source," https://www.dreamwidth.org/site/opensource. Accessed October 13, 2016.

67. Dreamwidth, "Business FAQs," https://dw-biz.dreamwidth.org/332.html. Accessed October 13, 2016.

68. For example, one person in attendance at a women in open technology unconference said she had learned to code out of necessity when no other players knew how to fix an online game she enjoyed (Fieldnotes, July 11, 2012, Washington, DC).

69. See Ellcessor 2016 for a brief discussion of Dreamwidth's efforts to be accountable to users who use assistive technologies.

> identity, and self-identification. We welcome activists, artists, bloggers, crafters, dilettantes, musicians, photographers, readers, writers, ordinary people, extraordinary people, and everyone in between. We welcome people who want to change the world, people who want to keep in touch with friends, people who want to make great art, and people who just need a break after work. We welcome fans, geeks, nerds, and pixel-stained technopeasant wretches. (We welcome Internet beginners who aren't sure what any of those terms refer to.) We welcome you no matter if the Internet was a household word by the time you started secondary school or whether you were already retired by the time the World Wide Web was invented. . . .
>
> Conservative or liberal, libertarian or socialist—we believe it's possible for people of all viewpoints and persuasions to come together and learn from each other. . . .
>
> To us, you're not eyeballs. You're not pageviews. You're not demographic groups. You're people.[70]

According to founders (one of whom is a woman), their effort to prioritize "people" had some curious effects on who joined the Dreamwidth volunteer base. Dreamwidth stands out as an open-source project welcoming to newcomers. As of 2012, 70 percent of people contibuting had never programmed or never programmed in Perl (the programming language in which the project is written) before coming to work on Dreamwidth, and most contributors were women.[71] (Needless to say, this was in stark contrast to many open-source projects, where 5 percent might be a high rate for women contributors.) To give a sense of scale, there were around 60 to 70 volunteers, and the project had around 40 to 50 *commits* [saves of changes to code] per week.[72] One person commented, "My favorite aspect of Dreamwidth is that every contribution is welcomed, even if it's incomplete or flawed. There is a sense that we want to help developers improve instead of rejecting them for not meeting some sort of quality standard."[73] According to founders and volunteers, it provides participants with a sense of pride in their

70. Dreamwidth, "Diversity," http://www.dreamwidth.org/legal/diversity. Accessed October 13, 2016. Earlier versions of this page in the internet archive Wayback Machine reveal that most of this text appeared in May 2009; subsequent modifications added such categories as "neurotype" and the sentence about age.

71. Fieldnotes, July 11, 2012, Washington, DC.

72. Smith and Paolucci, n.d.

73. Smith and Paolucci, n.d.

contributions to the project (even if they are small or relatively novice level), affective attachment to the project and to other volunteers, and opportunity for learning and growth in coding.

Dreamwidth is far from an enormous project, and it is barely a decade old. But it reflects many of the values of both open source and the diversity advocates within open technology. Running on open-source software, with its ambitions of limited growth and its profit structure designed to cover costs and feed revenues back into the project, Dreamwidth is a tangible example of what the Montréal activist was talking about when she commented that "we have more autonomy in software than in hardware or infrastructure."

Conclusions: Crafting Capacity?

This chapter has surveyed a variety of products that flowed from open-technology communities that explicitly prized and strove for diversity. Some of them, like the cryptodance and *Compulsive Repurpose*, were essentially speculative. Others, like feminist servers, represented functional objects that also played with and reimagined the underpinnings of technologies like networked computing. Still others, like Dreamwidth and the Lilypad Arduino, were not only concrete but sufficiently developed to be monetized and circulating in markets, albeit niche ones. All are of great importance for their symbolic or ritual dimensions. As these objects and ideas about them circulated within open technology and maker communities, they signified creators' beliefs about whose participation should count or be valued. In particular, elevating "feminine" craft products and practices and inserting them into spaces of electronics and computing enthusiasm should be understood as ritual communication, expressions of in-group solidarity and flag-planting.

Rather than focusing on the material effects of any of these products in particular, this chapter argues that collectively, they function as engines of sociality. In other words, they drive the formation of collectivities in open technology, offline and online. They form rallying points that exalt communal action, uniting an imaginary of hacking with diversity advocates' critiques of mainstream open technology. Participants make and remake such tangible artifacts as textiles embedded with soft circuits, open-source platforms for creative expression, and zines. These artifacts provide symbolic support for the evolution of nonhegemonic forms of hacking, around which communities and identities may be formed. This is true even when the material outcomes are far from working.

These products and artifacts also signified these communities' values. Indeed, they cannot be separated from the political orientations and emphasis on criticality formulated by diversity advocates. One prominent axis of critique, elaborated in a variety of ways, had to do with autonomy of infrastructure. One might reasonably ask, autonomy from what? And toward what end?

The hopeful Montréal hardware activist group quite ambitiously aspired to create a local supply chain for electronics hardware that relied upon reuse and repair. This represented autonomy from global supply and waste chains. Hardware was donated by local organizations and individuals, sometimes activists' employers (like the early feminist server). Dreamwidth, the software project, had carved out a niche using paid membership and mostly donated developers' time, so it was able to sustain itself economically without dependence on advertisers. In its commitment to free expression, it also supported the development of free culture products including fan fiction and artworks. The aspiring feminist hackerspace and the makerspace led by people of color also invoked autonomy. The former hoped to carve out permanent meeting spaces, sometimes borrowing space from employers or from other hacker groups. The latter articulated a skill set for surviving in uncertain times—a fusion of DIY, local and indigenous knowledge, and expert practices with everyday tools.

While these groups' aspirations to autonomy can be taken at face value, it is worth interrogating some of the evident paradoxes in these stances. First, technical expertise presents a difficulty for those simultaneously promoting rarified technical practices and a challenge to elitism in hacking. Quite simply, there is a barrier to entry with hacking hardware and software (as well as with sewing and crafting—but many people, especially women, are likely to have overcome that barrier early in their lives). As one activist at the Montréal feminist hacker convergence responded to the aspiring hardware repair collective, "It's not empowering to ask people to use Linux, or open source [more generally] if you don't have the support [in learning to use it]."[74] She herself was an enthusiast of free software. But she acknowledged that hardware and software autonomy might remain the province—and wish fulfilment—of geeks, not everyday users, unless the hardware collective gave close and considered attention to the politics of expertise, social position,

74. Fieldnotes, August 20, 2016, Montréal, Canada.

and difference. (The aspiring hardware collective members agreed. One of their goals was a form of mutualism; they pledged to help fix anyone's computer as long as it ran on Linux.) Dreamwidth's founders also recognized that expertise was an issue for a group building a technical product that claimed to welcome newcomers to Perl and to programming more generally; they emphasized that they chose to privilege newbies' contributions over an already-expert developer base: "Every contribution is welcomed, even if it's incomplete or flawed."

Another main paradox of the notion of autonomy of infrastructure is that diversity advocates are often required to eke out their communities' existences in impermanent, contingent spaces. At first glance, Dreamwidth's autonomy appears more secure. Colocation and temporal synchrony are not required for software collaboration; volunteer developers can be anywhere. The ethos of free and open-source software has allowed for the forking of even mature code bases, which allowed Dreamwidth creators to build upon and modify LiveJournal's code to suit their own purposes. But like other free software projects, in relying on enthusiastic volunteer developers, the software project chose a relationship to remunerated acts of production in which it is indeed dependent on paid labor, but paid labor that provides leisure or volunteer time on the side. In all cases, projects stake their autonomous existences in relation to paid work, more and less self-consciously (and sometimes including employer or corporate munificence in the form of surplus or waste).

In any event, as the feminist server authors remind us, autonomy as taken up here is generally more about *deciding one's own dependencies* than about a claim to total autonomy, or a fantasy of total control. Though many open-technology diversity advocates share with digital utopianists a belief in the emancipatory potential of computers, they position themselves differently than arch-cyberlibertarians.[75] These are not sites where it is universally believed that "code will save the world." [76] The universalism of that belief is belied in part by the fact that the communities invested in saving the world through code fall far short of representing the actual range of people in the world, a shortcoming that is emphatically certified by diversity advocates. In these sites, it is acknowledged that other forms of care and sociality can be as needed, or more needed, than code. Their border of care exceeds objects. This belief is expressed in part through the creation of products that are not

75. Golumbia 2013.

76. Golumbia 2013.

code, many of which have little obvious utility, and which refute and challenge universalist projects. A page for a zine or a handsewn LED on a sweatshirt (in 2016 in North America) signifies craft, the particular, imperfect, and nonscalable production. The feminist server is a particularly delectable refutation of universalism: it is situated, and it builds on the materiality of the bodies, hardware, and software gathered around it. Its creators have probably not only read Donna Haraway's "A Cyborg Manifesto," but also gone to sleep with it under their pillows for many nights in a row.

In the end, the products of diversity advocacy reveal a cultural project to intervene into hacking culture, which, like hacking culture itself, is rife with contradictions and an often underspecified, prefigurative politics. A critical stance is detectable, but not a full-throated political intention or intervention. This is perhaps unsurprising. As Christopher Kelty has argued, open source has become dominated by "domesticated" forms—its dominant impact has been one of "techno-infrastructural reform," less than a "public-oriented, critical and politicized" result.[77] The social and political milieu from which the diversity advocates' interventions and products emerge is not politicized in any clear way, nor is it monolithic (see chapter 6).

The above products are engines for sociality. They tell stories about belonging, which challenge the hegemony of hacking as an equivalent to coding and computer hardware. They also publicly critique the constitution of open-technology cultures.[78] This sociality may aid in the cultivation of networks of care that support people in their jobs or in other less-welcoming hobbyist technical spaces. Further political potential remains underdeveloped. Inarguably, this is in part because technologists convene around artifactual production. They often feel that dealing with technology offers a more concrete site in which to negotiate power and privilege—that is, it can seem cleaner to work on technological solutions than to wade into explicit social contestations. Feminist hackers in particular acknowledge that technological hacks cannot route around social problems, because they are products of social relations—though material artifacts are also produced. We are left wondering, is the production of *alternate stuff* that represents their values (or merely their presence) a sufficient outcome? Alternately, where else might the production of shared sociality and identity lead?

77. Kelty 2013.

78. Or, as Daniela Rosner and Sarah Fox write, "By claiming this labor as part of hacking cultures, the hackerspace members we discuss locate women's work at the center of new media industries" (2016: 566).

5

Working Imaginaries

"FREEDOM FROM JOBS" OR LEARNING TO LOVE TO LABOR?

On an electronic mailing list for Python enthusiasts, one person wrote in 2011,

> I just want to see more of a diverse crowd of hobbyists/enthusiasts playing with electronics, writing code for fun, and discovering that they can use these skills to actually make a living, like so many people currently do. . . . I've tried over the past seven years, running several women-only development efforts, but have not been able to sustain such a group for longer than a year. The response is generally: "Well, this was a fun, great resume building exercise. Now I have to get back to real life. Bye!"[1]

Her quote deserves special attention for the multiple issues at stake in focusing on diversity as a primary concern in her software community. First, we note that she invokes "a diverse crowd," but goes on to refer specifically to "women-only" development efforts. Second, the poster writes that development "is fun" and suitable for "hobbyists/enthusiasts" interested in "playing with electronics." Last, she laments that the projects she has spearheaded have tended to founder, suggesting that participants may have viewed them

1. Email,—to [Python list], February 2, 2011.

as mere "résumé-building exercises." Her disappointment seems puzzling when we consider that earlier in her post, she writes, "they can use these skills to actually make a living."

As Sara Ahmed writes, the "mobility of the word 'diversity' means that it is unclear what 'diversity' is doing, even when it is understood as a figure of speech."[2] The above diversity advocate's quote underscores Ahmed's point, that even though we understand what *diversity* means, it is not clear what work it is doing, or is meant to do. In her statement, the mobility and polysemy of the word gives rise to ambiguity and tensions over *who* should participate in the development and production of electronics and software, as well as *why* participation in these pursuits is meaningful to participants, or within the wider society.

Technology is a unique domain for the discharge of political energies. In the popular imagination, it has been vested with the power to initiate change.[3] Many technologists, especially those in activist geek circles, are specifically motivated by political concerns and seek to build technologies that they believe can redress social imbalances or inequities. The impulses to open up technological participation span a range of political motivations. At one end of the spectrum is something akin to Sandra Harding's refutation of "value-free" science, which shifts the burden of science away from neutrality and toward a starting point that takes into consideration the needs of disenfranchised groups;[4] by extension, democratized technology absolutely requires alternative perspectives and practitioners. At the same time, because diversity is such a protean concept, it can easily shade away from this more radical stance and into very different political valences. Another salient rationale that diversity advocates offer for their efforts focuses on labor, production, and consumption; they invoke diversity as a good in market relations. An extremely common argument for expanding the pool of technologists is that this will result in the entry of underrepresented groups into the workplace, which will change what gets built, which will in turn capture (or serve, depending on the perspective) a wider consumer market. These motivations for diversity work are potentially divergent at their cores, but sit together well enough that both are consistently found in diversity advocacy around open technology. One should not, however, be mistaken for the other.

2. Ahmed 2012: 58.
3. Marx 2010: 577.
4. Harding 2016.

It is significant that the events chronicled in this book occurred in the aftermath of the worst financial crisis the United States had experienced in seven decades (the so-called Great Recession precipitated by the collapse of the US housing market in 2007 to 2008). Anxiety over employment, job security, and wages ran high for many people. Boosters promoted work in IT fields, holding it as a growth area with a high potential for gratifying remuneration. The reality is much more complicated, because how the boundaries of IT work are drawn starkly demonstrates that some work with computers is higher status and higher paid (and thus vaunted), while other work is lower status and lower paid (and thus often rendered invisible). Indeed, the prevailing understanding of IT work as high status has blind spots surrounding class, race, and gender, both domestically and transnationally.[5] As Virginia Eubanks has shown, workers who hold jobs in data entry, telemarketing, call centers, and the like are not rewarded with high status nor with desirable employment conditions.[6] Nonetheless, computer competency has long been held out as a critical component of employability and social standing, which intensified in the wake of financial calamity. That economic climate and the dominant understanding of IT work as high status are crucial context for the discussion that follows.

A salient strand of discourse about diversity in open tech, which is of special interest in this chapter, concerns the workplace. Participants commonly hold that diversity is important because it empowers members of underrepresented groups to claim jobs in technical fields. This can be equated with social equity, but it is also consistent with companies' desires to capture a diverse consumer market; many hold that having a wide range of people in product development is conducive to courting a wide consumer base. Others de-emphasize job readiness, wishing to see voluntaristic technical communities composed of a diverse range of participants for more inchoate reasons, generally framed as strengthening FLOSS and an attendant movement through pluralism. Still others undertake collectivity formation around open technology as an expression of radical politics, hoping that diversity work can serve as a mode of intervention into and critique of the

5. McCall 2001; Poster and Wilson 2008. And of course higher-status tech work is supported by the labor of poor women and women from minoritized racial and ethnic backgrounds who work as administrative assistants, food service, and custodial staff, as well as childcare providers and housekeepers. Lower-wage contract workers at tech firms are disproportionately African American or Latinx (Young 2017).

6. Eubanks 2012: 72.

dominant social order, including questioning capitalism and formulating alternatives to it. This chapter zeroes in on ideations surrounding work and labor relationships within diversity initiatives and demonstrates that various motivations for diversity advocacy sit in tension with one another. It argues that the imagined relationships between diversity in tech and workplace preparedness are important because they expose the generative potentials in diversity advocacy.

Politics of Open-Technology Projects and Their Relationship to Paid Labor

As noted in this book's introduction, many scholars of hacking and tinkering have focused on the fact that these activities often take on meaning as communal and shared actions, even while often denying formal politics except for software freedom.[7] This rejection of formal politics has made FLOSS an unlikely site for gender activism, at least historically.[8] But FLOSS projects are heterogeneous and have matured over time. They are also in dialogue with the wider culture, which is currently awash with calls for diversity in tech, especially in relation to industry and education.

The relationship between paid work and open-technology projects contains many somewhat paradoxical elements. As STS scholar Anita Chan writes, "Few practices seemed to be so effective at generating the intense enthusiasm and heightened investments of global free labor that free software participants—as highly skilled information classes, no less—so extravagantly displayed."[9] As has been amply explored elsewhere, many FLOSS projects depend on volunteered time (time that is subsidized, on some level, through paid employment that generates "leisure time"). This free labor feeds back into the market in multiple ways, creating value for coding languages, FLOSS projects, and the like. Such voluntaristic participation can also benefit individual participants by increasing skills and reputational capital that may be useful when seeking paid employment. And paid workplaces may encourage participation in labor-of-love projects and pursuits. Google, for example, has allowed its developers to split their paid time on an 80/20 basis, working on assigned Google work 80 percent of the time and on their

7. Coleman 2012a; Dunbar-Hester 2014.

8. Nafus 2012; Reagle 2013.

9. Chan 2013: 117.

own "passion projects" the other 20 percent; communication scholar Fred Turner argues that this is a productive strategy for the company, which sees both an enhanced product line and motivated employees.[10] While this chapter does not contribute to debates surrounding FLOSS participants as free labor for the IT industry, these discussions are necessary background for the sites under consideration.

The history of these relationships is no less free of contradiction. According to George Dafermos and Johan Söderberg, "there is [a] line running from the white-collar engineers of the 1950s to the present-day hackers; [on the other hand,] another line connects hacking to the resistance of the machine operators working under those engineers."[11] For a variety of reasons, including programming's tendency to flicker between craft and science, programmers have long struggled with issues of autonomy and managerial control.[12] Job precarity in the symbolic epicenter of the IT industry, Silicon Valley, is legion and has been for decades.[13] This insecurity occurs in spite of an often-invoked notion of programming as high-status work.

Lastly, amateur pursuits around electronics have a long history not only as sites of affective pleasure but as sites where skills and an affective attachment to technology are learned that are additionally useful in paid employment in technical fields. Such practices were stable through much of the twentieth century, as documented by historians of radio.[14] Men's and boys' leisure activities allowed masculinity to refashion itself, transforming mastery over rugged nature to mastery over electronics; brains outstripped brawn in a modern, technological masculinity.[15] Early computer workers and hobbyists often tinkered with radio electronics as young men. Even before hobbyist and PC computing became prevalent, Ensmenger argues that established labor practices reinforced men's primacy in computing. In workplaces, in keeping with propriety, women were sent home after business hours ended to keep them from mingling with male employees, giving men exclusive mainframe access on weekends and in evenings. This had the effect of showing up the association of computing with masculinity, even though computing work was in actuality conducted by both men and women.[16]

10. Turner 2009.
11. Dafermos and Söderberg 2009: 56.
12. Ensmenger 2010b.
13. Turner 2009: 77.
14. Douglas 1987; Haring 2006; see also Marvin 1988; Oldenziel 1997.
15. Douglas 1987, chapter 6.
16. Ensmenger 2010b.

Working Imaginaries along a Continuum

I now turn to the imaginary of paid work within these sites of voluntarism and advocacy. What is curious is that there is enough of a shared social imaginary to bring people together in these common spaces, but there is little coherence regarding the ideation of work. The shared ideation has to do first with technology as a worthwhile enterprise, and second with a notion of social change, especially in terms of challenging technical cultures as sites of social exclusion.

Diversity initiatives in open-technology projects represent a wide range of impulses, practices, and goals. This chapter aims to take seriously the heterogeneity of diversity-centered efforts. It zeroes in on the imagined relationship between employment and open-tech diversity efforts in order to evaluate these efforts' generation of value and values. Rather than focusing on outcomes or outputs, here I examine the motivations and imaginaries of diversity advocates. It is apparent that ideations drive participation, and that the factors that motivate diversity work in these contexts are neither monolithic nor entirely straightforward. What exactly do advocates hope to change, and why? What is the relationship between envisioned change and work or labor relationships?

At first blush, the answer to these questions is underspecified. One typical practice is illustrated in figure 5.1. The photo depicts a workshop held over one evening and one weekend day (hours when students or people with "regular" work hours are not at workplaces or in school, though of course domestic duties do not cease), geared toward women who were interested in learning to program in Python. ("This workshop will be a great way for you to explore both programming and Python," read an email from organizers.[17]) It was held in a university classroom in Philadelphia, PA, in June 2012, and attended by about twenty people.

This represents a wholly typical event within FLOSS diversity activism. In open-source spirit, the Python tutorial module was borrowed from a group in Boston, MA, which ran similar women-focused events there and shared the instructional template. It was taught by unpaid volunteers in a borrowed space. Around twenty people attended. Other than women "exploring programming and Python," no explicit goals for the workshop were set.

At another event that summer, a two-day unconference dedicated to "increasing women's participation in open technology and culture" held in

17. Email,—to—, June 21, 2012.

FIGURE 5.1. Python tutorial geared toward women programming novices, Philadelphia, PA, June 2012. Photograph by the author.

Washington, DC, around two hundred participants organized into various sessions based on their interests. One session was titled, "Kill your boss and take his job: A plan for world domination, or, how to move up the ladder" (figure 5.2).

Though of course this session title is tongue-in-cheek, it overtly references workplace relations, specifically women's ascension of the corporate ladder and efforts to wrest control from men managers. Its reference to

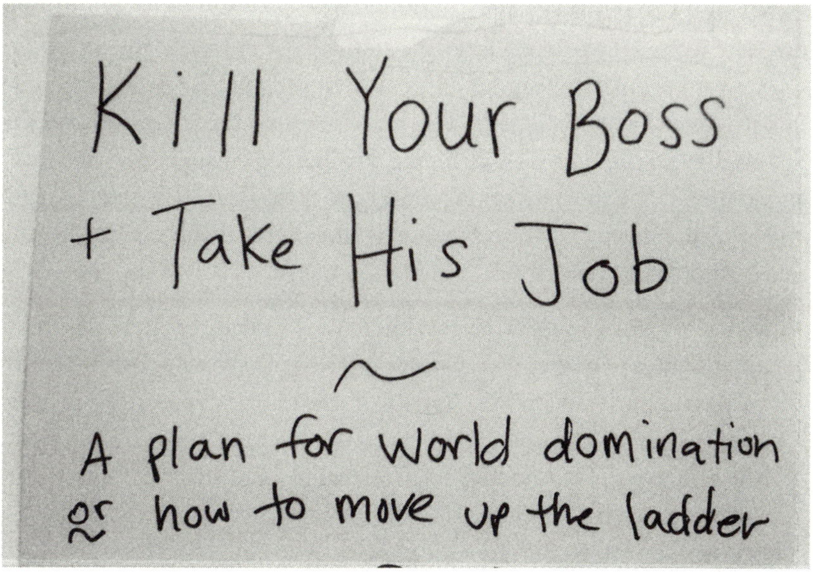

FIGURE 5.2. Session proposal, "Kill your boss and take his job," unconference for women in open tech, Washington, DC, July 2012. Photograph by the author.

"world domination" is humorous, but it notably circumscribes its reimagining of power relations as *having to do with workplace relations.*

And this goal of workplace advancement was reinforced many times: in a post-unconference email, one of the organizers wrote to all of the attendees, "Another thought from dinner: a jobhunting and career planning session. Not a theoretical 'how women hunt for jobs' session, [but] a practical session you go to if you are jobhunting or hiring right now, or want to help women get a job. Activities are things like 'oh, I could find out for you if my friend at COMPANY knows of any openings.' So, um, I guess they call that networking :) ".[18] Another example is this email sent to a mailing list in 2015, where a poster wrote, "a few months ago, I think [Name] had this idea for a workshop where people critique each other's resumes?"[19] Repeatedly, both unconference attendees and organizers reinforced the notion that a primary motivation for their diversity advocacy was to *advance women's standing in the workplace.* Networking and résumé critique were proposed as appropriate activities for collective engagement to follow in-person unconference

18. Email,—to [Women in Open Tech List], July 10, 2012.
19. Email,—to [Women in Open Tech List], March 23, 2015.

gatherings. Labor anxiety and aspiration drives much of the advocacy for diversity in tech, even in voluntaristic settings.

At the same time, other goals and values are commonly expressed around diversity advocacy. Liane said in an interview that the social critique she believed in ran deeper than changing the gender balance in IT industry employment. She spoke of her personal history that had led her out of high-status programming work and into full-time diversity advocacy. She said that hackers and coders have

> faith in progress, science and technology. . . . People want to do something with a purpose, that has a point. I think people want to believe they're doing a good thing. [But now I see it as,] if you're defining [progress] as building a better product that's concentrating wealth for a few, [the status quo] is working great, [but this assumption is wrong]. When I started [as a programmer] I was doing capitalism [and I was fine with that, but] I no longer make the argument about [building] a better product. There is a collision between science kids with a nerdy mindset who want to do good, who work for an industry that is corrupt.[20]

She said that her thinking had changed over time, and she had evolved from someone who was gratified by puzzle-solving in coding work and even the transcendent ideas of open source, in which good code will be put to use over and over, into someone who no longer felt that solving technical problems or "building better products" would, in itself, induce social positive change. But she felt that she needed to tread lightly in the tech industry regarding her choice to no longer "do capitalism"; for example, in our interview she insisted on speaking to me "as herself," expressing personal beliefs, and not as someone affiliated with her foundation that was working to increase women's participation in open technology. Clara, a participant in a feminist hackerspace, told me that she wondered about the degree to which the "jobs for women in IT" rhetoric could be cover for more radical activist goals, a discursive Trojan horse that could attract funding and allow activists to build institutions around their values.[21]

Preceding another unconference planning meeting in 2015 (which was scheduled but did not occur), one would-be attendee wrote to the planning list to inquire whether someone might lead a session on worker-owned

20. Interview, Liane, July 24, 2014, San Francisco, CA.

21. Personal conversation, March 3, 2011, New York, NY.

cooperatives and other less common business structures.[22] She wrote, "I've been hearing a lot of feminists saying they'd like to work for a feminist/women-friendly worker-owned tech coop and I know there are some attendees who know about this!"[23] Her comment illustrates the DIY nature of the unconference, where topics of interest are floated and then molded to fit collective desire. Because this session never happened, I obviously cannot report its content. But it is worth noting that for some people, the concept of worker-owned co-ops was a natural fit for this meeting. It raises the possibility that alternative business models could be one possible outcome of diversity advocacy, though the chatter in the course of planning a session that never occurred is a long way off from a living, breathing worker-owned co-op, let alone a transformation of capitalistic relations on a wider scale. In addition, it would be hard to claim that worker-owned co-ops were a central concern here; relative to the amount of traffic on this list and other similar ones about career networking, this topic was a mere trace.

In a similar vein, diversity advocates at the two unconferences I did attend devoted much energy to discussing *impostor syndrome*: "the sense that everything you have accomplished is fraudulent and you are constantly about to be found out."[24] Multiple sessions across the two unconferences I attended addressed impostor syndrome, and the initiative that housed them left behind a guide to impostor syndrome training even after it was shuttered. At the 2012 Washington, DC, unconference, one session I attended included discussion of how people felt like they were impostors, and offered practical tips for claiming more authority in the face of these feelings. The person leading the session explained that she had experience with these issues, as she ran an open-source platform for sharing content created by users, particularly fan culture projects. She said that 70 percent of the people who had created the platform had never programmed before the project, and women were a majority of the people who built and maintained the platform, so she had had moments where she thought, "Oh my god, how do they let us have a business?"[25] She offered a number of suggestions, particularly around how women were likely to use qualifying language when describing what

22. See Costanza-Chock 2019; Scholz and Schneider 2017.

23. Email,—to [Women in Open Tech List], March 17, 2015.

24. Paolucci 2013.

25. Fieldnotes, July 11, 2012, Washington, DC. The platform offers paid accounts to users who desired premium features, which is how it sustained itself; its organizers rejected advertising as a revenue source. It was organized as a for-profit company that paid a couple of employees and operated in some ways like a nonprofit (two-thirds of profits were pledged to go back into

they did or what they knew. She said one way to evaluate and change these patterns was to, for example, write an email explaining a technical issue or problem and then go back and delete all the language where one was being mealy-mouthed about one's expertise: "go ahead and write the email, and then go take out all the qualifiers, [like] 'I don't know if this is right, but . . .'" One of the people in the session said she had come to coding from fandom because nobody in her gamer community knew how to fix some aspects of the code in the game that she and fellow players found unsatisfactory. She said that she then took computer science classes and felt like she "didn't know anything but wound up tutoring half the class."[26] Everyone in the session appeared to relate to these feelings of inadequacy and shared stories and advice. Given the high degree of interest in these sessions among unconference attendees in both Washington, DC, and San Francisco the next year, it seemed that the sessions were serving a purpose, if only to remind people that these feelings of insufficiency or fraudulence were fairly common (the organizer of the DC session said, "guys are vulnerable to it too"). These sessions "promoted an intimate and supportive environment that made it easy for women to reach personal insights," to borrow Francesca Polletta's words about consciousness-raising rap sessions in the women's movement of the 1960s and '70s.[27]

It is challenging to think through these limits and possibilities. One person who had recently led the Boston group's Python tutorial for novice women in her city, Columbus, OH, reported back on her experience via email. She said:

> At the Columbus Python Workshop, some of the students and I talked about how we could use workshops for not just gender diversity, but economic diversity. But [this raises lots of questions] . . .
>
> —Is this really a practical skill for people in tough economic straits? Obviously programming is a great career; just as obviously, nobody's ready to start at Google the Monday after a weekend workshop. The workshop is obviously not just about careers—it's also about having fun, building self-confidence,

the community, according to its "Business FAQs" page), so while it was sustainable, it was not wildly profitable.

26. Fieldnotes, July 11, 2012, Washington, DC.

27. Polletta 2002: 164.

understanding our computerized world better, etc.—but I don't know if those noneconomic motives will ring hollow.
—And if it is practical, how can I *sell* its practicality to get agencies onboard and students in the door?[28]

She reflected on how the workshop offered "fun, self-confidence, understanding our computerized world better" to attendees, but noted that to sell it to would-be funders and many attendees, economic benefits would need to be emphasized. At the same time, she correctly pointed out that to actually reap economic benefit, novices to programming would need to put in much more time and effort than a weekend workshop. (As an utterly novice programmer, I can attest that the similar event I attended as participant-observer in Philadelphia left me a very long way from attaining even basic working proficiency in Python, let alone contributory expertise.)

At the Philadelphia hackerspace, HackMake, members reflected on issues of economic opportunity, in their space and more widely. As noted in chapter 3, in a discussion of diversity and outreach, one person observed that the space's current membership was fairly homogeneous. He said, "It's no coincidence [that current members] were looking for [this community, and] we found what we were looking for [at HackMake] and we all happen to be that 'type.' Call it primary requirements; geek, engineery, and employed."[29] This statement highlights a few important issues. Elsewhere I have remarked upon how "geek" and "engineery" implicitly mark social identity including gender and, to an extent, race. What is of interest here is how the hackerspace member also notes that current members tend to be employed, which, along with "geek" and "engineery," marks class. This exposes hackerspaces as a contemporary iteration of the long-standing connection between technical employment (being "engineery") and technical pursuits in leisure time. While whiling away hours in a hackerspace is not necessarily reducible to a hobby (and many would oppose this characterization), it appeals most to people with technical affinity and some spare time to burn—and in the Philadelphia hackerspace this meant people who were comfortably enough employed that hackerspace dues and finding spare time to tinker or hack were not burdens.

Other members took the opportunity of this discussion of the hackerspace's homogeneity to explicitly draw connections between the relatively

28. Email,—to [Python list], January 26, 2013.
29. Email,—to [HackMake list], February 5, 2011.

resource-rich hackerspace members, the climate of financial insecurity, diversity concerns, and an inchoate goal of serving or reaching out to the wider community. One person wrote,

> We have a rare opportunity now with the economy being what it is to offer [classes] that might otherwise not garner an audience. Such topics as How to Change Your Own RAM, Photoshop 101, Basics of Web Design or even How to Blog . . . might do very well if peppered among our standard offerings. This would also address any concerns that we are becoming too [homogeneous]. In short, it's an opportunity for us to offer instruction on topics that may help people get back to work while introducing them to [HackMake] in a very non-intimidating way. Hopefully by doing this we will also gain new membership and new perspectives as well.[30]

In other words, he proposed conjoining the diversity mission of the hackerspace to the goal of "helping people get back to work," supposing that opening up hackerspace pedagogy to job-relevant classes might push its membership past its current core. This is a rosy picture, but like the person teaching Python in Columbus, quoted above, one wonders how realistic or desirable it would be for unemployed or underemployed people to spend time in the hackerspace learning skills that would possibly translate into employment opportunities. It is also unclear whether the hackerspace would be equipped to offer vocational training to the degree needed by people who were novices; a one-off class on installing RAM or Photoshop software might be interesting and useful to someone who was otherwise motivated to learn but would be unlikely to stand in for deeper training. It also assumes that people have working computers and software at home, and time and motivation to deepen their expertise, which is certainly true for some, but not for all in "tough economic straits."

In any case, however well intentioned, these voluntaristic urges to reach out to people without jobs and enact job training through DIY software and hackerspace workshops may fall short in critical ways. The exhortation that various groups underrepresented in tech learn to code (or solder, or install RAM) in order to improve their social position burdens *individuals* with the onus to bootstrap or lean in. Furthermore, this exhortation draws attention away from social and economic policies that contribute to members of certain social groups occupying more marginalized positions in the

30. Email,—to [HackMake list], February 4, 2011.

first place, and places the burden on people who may already be afflicted by precarity and structural inequities. Improving the stations of unemployed or underemployed people in their communities is a noble goal, but it is not clear that voluntaristic DIY organizations are equipped with the right tools to make meaningful interventions; the right tools may exceed both programming exercises and the forms of social support offered by a volunteer group of technophiles. (It is also curious that voluntaristic organizations would believe their missions include job training for people not already active in their organizations. In past eras, employers were often responsible for providing job training for entry level IT jobs,[31] not leaving volunteers to create economic value for their industries. As Melissa Gregg writes about civic hackathons, "civic hackers 'learning to labour' . . . in informal settings create an industrial reliance on donated work in the process of receiving professional training."[32])

A HackMake board member praised this suggestion to offer basic classes with potential vocational application but parsed the issues slightly differently. She replied,

> As you might imagine, I'm utterly into this idea. . . . Democratizing and demystifying technology . . . is one of my favorite things to do. . . .
>
> Previous experience tells me that people absolutely GLOW when they learn how to make cables. Any time I've taught RJ45 [data line connector] crimping or soldering of audio cables, the result is almost rapturous. These money-saving ideas are about 10 degrees off from the back-to-work skills that you're talking about, but it's in the spirit of getting people past technology barriers and into the fun stuff. That moment when people realize that the tech in front of them isn't that hard, and if it isn't that hard then maybe there's all kinds of other tech that's within their reach, well, that moment is magic and I'm eager to help more people experience it.[33]

Though she supported the idea of doing diversity work for the hackerspace by offering new classes to draw in different sorts of folks, her emphasis was less on vocational training and more on demystification, helping people "get past barriers" and *into a realm of practice where joy and agency are palpable.* This is also held to be transformative, but less instrumental. And it reaffirms that one of the reasons people engage with technology is for their

31. See Abbate 2012; Downey 2002, chapter 8; Ensmenger 2010a; Miltner 2019.

32. Gregg 2015: 193.

33. Email,—to [HackMake list], February 4, 2011. Emphasis in original.

own pleasure and curiosity. This affective connection to technology and to other people through those shared pleasures are what undergird the communal aspects of voluntaristic technical pursuits. She directs the conversation to expanding the hackerspace's membership by expanding technical affinity—helping more people to "glow" over their exploration and creation of technology. (She does mention the "money-saving" aspect of DIY, i.e., that making cables could be cheaper than buying them, but this is probably not a primary motivation. Also, of course, much DIY requires consumption to participate.)

By contrast, other diversity advocates have elected to address diversity issues in ways that are explicitly at odds with generating value for the IT industry within prevailing social and economic relations. In an announcement of a 2016 event in Montréal centering on diversity in tech, organizers wrote, "The event aims at addressing the lack of women, queer, trans [people,] and diversity in technological fields in general and hacking more specifically. But even more so, it aims at creating a community that critically assesses the hegemonic narratives around technologies, the modern[ist] aspects of its underlining Western assumptions and its inherent capitalist inflections, among others."[34] Along similar lines, the Oakland makerspace led by people of color listed the following during a discussion of values and vision for the space: "[to be a] welcoming place for poor women, trans, low income, formerly incarcerated . . . of and for [the] immediate community . . . political education, social justice . . . *freedom from jobs*."[35] In both of these examples, advocates for diversity articulate an alternative value system they hope to implement around technology and technical practice, wherein not only *who* participates in technological production is changed, but *why and how* people engage with technology is altered.

Advocates critique how capitalism has shaped hacking and making, and present the underspecified but provocative notion that a makerspace can provide "freedom from jobs," as opposed to conscripting members of marginalized groups (such as members of minoritized racial and ethnic groups, low-income people, and formerly incarcerated people) to be better producing and consuming capitalist subjects. Critics such as Angela McRobbie have noted that in the knowledge work and creative fields, "work has been re-invented to . . . become a fulfilling mark of self." These makerspace participants reject what McRobbie characterizes as an "attempt to

34. Email,—to [Women in Open Tech List], June 1, 2016.

35. Fieldnotes, July 24, 2014, Oakland, CA; emphasis added.

TABLE 5.1. U.S. Labor Data, 2011–2017

	Computer and Mathematical Occupations		Other Computer-Related Work		Service Work	
	Average Wage	Workforce %	Average Wage	Workforce %	Average Wage	Workforce %
2011	$78,730	2.656	$35,640	0.314	$23,025	11.567
2012	$80,180	2.746	$36,043	0.3	$22,965	11.787
2013	$82,010	2.788	$35,880	0.285	$23,145	11.993
2014	$83,970	2.837	$37,693	0.265	$23,480	12.16
2015	$86,170	2.905	$37,682	0.436	$24,250	12.244
2016	$87,880	2.967	$39,968	0.229	$25,180	12.462
2017	$89,810	2.99	$41,023	0.209	$25,990	12.874

Source: Occupational Employment Statistics Program, Bureau of Labor Statistics, US Department of Labor. *Note*: This table shows national US labor data in several sectors during the period of Columbus, OH, and Philadelphia, PA, diversity advocates' reflection on whether their activities contributed to workforce preparedness. Note that service is the sector that exhibits the most growth, and that in IT, status and pay varies considerably, for example between being a systems administrator and holding a data-entry position. Lower-status service work often incorporates IT in order to be performed remotely—thus blurring the boundaries of whether it is service or IT work—and is often offshored (Cameron 2000; Freeman 2000; Poster 2007). A prominent example is call center work. Content moderation, though not customer-facing service work, is also an example of low-status IT work that includes significant emotional labor.

make-over the world of work into something closer to a life of enthusiasm and enjoyment."[36] The makerspace is understood not as a training ground for work, but as a place to *experience making as both politicized and distinct from capitalist production*; the makerspace members reclaim their enthusiasm and enjoyment for making from their work selves. These goals are consonant with Virginia Eubanks's claim that many low-income women and women who are members of minoritized racial and ethnic groups do not lack experience with IT, but do not experience IT work as empowering because for them it is often low-status, casualized, and heavily surveilled (see table 5.1).[37]

Now I return to the quote that opened this chapter:

> I just want to see more of a diverse crowd of hobbyists/enthusiasts playing with electronics, writing code for fun, and discovering that they can use these skills to actually make a living, like so many people currently do. . . . I've tried over the past seven years, running several women-only

36. McRobbie 2002: 521.

37. Eubanks 2012. See also Margolis 2010.

> development efforts, but have not been able to sustain such a group for longer than a year. The response is generally: "Well, this was a fun, great resume building exercise. Now I have to get back to real life. Bye!"

This post to the diversity email list neatly encapsulates many of the values of diversity advocacy, including those that essentially contradict one another. If the poster wanted the participants in her women-only development groups to mainly derive workplace-relevant training from her efforts, she would not decry inconsistency when they left the group having gained a line for their résumés. And yet she is palpably disappointed by this, even though she also sings the praises of writing code for its potential to help one make a living. Something is missing for her, though she does not say what.

Conclusions: Can Diversity Advocacy Do Scary Work?

This chapter has argued that the political motivations of diversity advocacy in open technology are ripe for analysis, and that the work imaginaries of advocates are a lens into the differing and sometimes conflicting impulses for opening up technical participation. Diversity advocacy represents a wide swath of activity and is flexible enough to encompass a variety of practices and discourses. Discourses about workplace empowerment are common enough to warrant sustained scrutiny.[38] While we can hardly fault people for pursuing job security or value as a worker, the wider emancipatory politics imagined by some who pursue and promote technical engagement is not consistently evident here.

What should we make of all this diversity within diversity work? On the one hand, some people enjoy hacking, making, and coding, and they wish to open that enjoyment up to others, in particular to create safer spaces for people who might be drawn to hacking or making but intimidated by hegemonic (masculine, white, elite) tech cultures. What people then choose to do with those skills—how they choose to enjoy them—is left relatively underspecified. That said, it is routinely acknowledged that it is discursively easy, and sometimes strategically useful, to align diversity in tech advocacy with industry goals and market values, nominally funneling programmers, makers, and hackers toward jobs in IT and using voluntaristic pursuits to suture leisure time to work and vice versa.

38. Discourses of empowerment for disadvantaged groups through coding also have a history; they are as old as the 1960s (Abbate 2018).

On the other hand, it is clear that diversity advocates intend outcomes beyond technical training for IT industries. Christopher Kelty has written that making has been given "an immediate, ethically inflected, political urgency. Furthermore it hails subjects who desire to engage (and take pleasure) in this kind of making: contemporary subjects of capitalism who want individual agency to be combined with . . . collective experience."[39] "Collective experience" is plainly part of what draws people to participate in the "transcendent" aspects of FLOSS, and to identify as makers in other contexts. A founder of a Seattle hackerspace that emphasized inclusion remarked, "Hacker spaces are a sort of gateway into exploring everything. By encouraging the taking apart of 'closed' objects . . . we can begin to form mindsets which make exploration and understanding necessary joys in life."[40] "Exploration" and "taking apart closed objects" are politically inchoate, but they potentially point to realignment of power relations, especially when experienced collectively.

And yet, as Kelty rightly argues, these attitudes are formed within and suffused by contemporary capitalism. This is explicitly acknowledged by the organizers of both the Montréal event and the Oakland makerspace, who articulate alternative formulations and challenge the relations of production that drive so much tech participation, even volunteer production. It is evident that market values, diversity advocacy, social justice, and other goals, including recuperation of nonmarket value, are not neatly separated. It is also apparent that whatever the goals of diversity advocates, some of the people drawn in by diversity initiatives see an obvious utility in the workplace-related aspects of diversity outreach, as noted in the "Kill your boss" session and résumé critique proposed by unconference volunteers.

Thus, while some of the rhetoric in diversity advocacy imagines the possibility of deep social critique, much reads as politically agnostic beyond a redistributive impulse that will bring people underrepresented in the IT industry into paid tech work. Why does any of this matter? We might reasonably say, "We don't much care if a bunch of do-good geeky people lack a clear or unanimous mental picture when they come together to widen the ranks of technical participation." And yet it matters because at the core of these efforts is a belief in technology as a site of political potential. Technologists,

39. Kelty 2013.
40. Davies 2009.

especially politicized technologists, often want to build tools to do good.[41] Instead, the vocabulary of political imaginary in diversity initiatives largely recapitulates notions of expanding economic opportunity and configuring underrepresented people as tech workers, especially for corporations that ultimately hope to capture a diverse consumer market.

This is a missed potential. Representation in particular as a goal has limits as a project of empowerment, as noted by scholars of postfeminism and race.[42] With regard to technology in particular, as Ron Eglash et al. argue, underrepresented groups' participation in technological production or consumption from which they had previously been excluded is not necessarily indicative of a change in social power or social status.[43] It is easy to conflate elite social power and technical participation, but they are not interchangeable.[44]

Diversity advocacy appears to bring together people whose politics and agendas might not otherwise align.[45] This diversity work could be viewed as building capacity and social infrastructure for sustained political challenge to prevailing technical cultures and industries. Another possibility, though, is that *diversity is part of the problem*: as Ahmed writes, diversity can easily "detach from scary issues, such as power and inequality."[46] Voluntarism centered on open technology geared toward diversity is potentially of a piece with market values. In itself, this voluntarism can do little to dismantle structural inequity, even at the level of rhetoric. Its technologies and practices would ideally be coupled with social justice movements, national policy changes, or other broad social forces in order to effect deeper social change, as Ron Eglash writes.[47]

To bring this point home, it is worth expanding on Eubanks's points about IT work among low-income women in the United States. Especially globally, IT work can hardly be said to have a diversity problem.[48] Much offshore IT work is *pink collar*: low status and performed by women.[49] This should immediately give rise to the recognition that rather than advocating

41. See Wisnioski 2012 for the history of struggles within engineering on how the field should or should not address matters of social conscience; see also Coleman 2017.

42. McRobbie 2008; Gray 2013.

43. Eglash et al. 2004: xv.

44. Dunbar-Hester 2014: 188.

45. Star and Griesemer 1989; Dunbar-Hester 2013.

46. Ahmed 2012: 66.

47. Eglash 2016.

48. Thanks to Paula Chakravartty for discussion on this point.

49. Freeman 2000; Huws 2001.

for more women in tech work, it would be more to appropriate to advocate for more high-status, well-remunerated work. More women in tech could easily be realized as more feminized labor, which is definitely not what advocates have in mind. And women in minoritized racial and ethnic groups would likely bear this burden disproportionately: research suggests that that racial inequality *among* women is more severe than gender inequality among whites.[50]

To call for more women in tech alone is to miss the point that "computer technology is pivotal in the neoliberal reformation of capitalism that most people have encountered as weaker unions, flexible labor markets, and deskilling."[51] And many of the practices of diversity advocates seem unconnected to these realities. While "impostor syndrome training" seemed to resonate emotionally with many attendees at both unconferences, it is far from clear how this practice attaches to meaningful changes in workplace dynamics, let alone broader social and economic structures that affect labor relations.

It is interesting to reflect on the cultural space of diversity in open tech, which can attract and hold people whose beliefs and aspirations may diverge so greatly. Open-technology voluntarism simultaneously produces capitalistic relations and a fantasy of social relations outside of capitalistic relations. This means that it can simultaneously support an imaginary in which capitalistic relations may be rewritten or destabilized and—more commonly—one in which they are reinforced. In order to maximize the potential for generation of social justice, not only redistribution, both *diversity* and *technology* need to be reexamined as platforms for social change. Using them as orienting concepts requires a stronger acknowledgment of their potential to maintain or reinscribe prevailing social and economic relations. While neither diversity nor technology is without redemptive and generative potential, they are often implicated in an impoverished politics where activism is cast in market-friendly terms. Advocates serious about generating justice must take on the hard work of hitching both diversity and technology to "scary" issues, such as power and structural inequity. If workplaces are of central concern, advocates may need to consider not only voluntaristic and local pipelines,

50. Eubanks 2009: 119, paraphrasing McCall 2001 (Eubanks's emphasis). Green argues that African American women's employment possibilities have often been disproportionately adversely affected by technological displacement (1995).

51. Dafermos and Söderberg 2009: 67. For an example, see De Lara's analysis of how labor, space, race, and class are constructed vis-à-vis computer tracking and logistics (2018). See also Downey 2001 for thoughtful discussion of the "where" of labor in information internetworks.

but also structural forces that exalt some workers at the expense of others. Tech work is not monolithic; rather, it is constituted of interrelated industries, with differential status and privileges. Devoting unapologetic care to social structure and inequality (scary though these issues might be), how might advocates reframe the central, yet ambiguous role of workplaces in their analyses? Though tech work sparks their intervention, a rigorous analysis begs for the inclusion of a broader spectrum of workplace relations. "Entering the makerspace as a janitor"[52] should be taken seriously, at face value.

52. Chua 2015; Irani 2015.

6

The Conscience of a (Feminist) Hacker

POLITICAL STANCES WITHIN DIVERSITY ADVOCACY

During the June 2013 San Francisco unconference for women in open technology, news broke about a massive leak regarding warrantless National Security Agency (NSA) spying. Only a few days earlier, on June 5 and 6, the *Guardian* published its first articles showing that US telecommunications companies like Verizon had been handing over user data to the US government and that other tech companies like Google and Facebook had given the NSA direct access to their systems. The immediate reaction by the US government was to defend these programs, with President Barack Obama saying they were correctly and legally overseen by the courts and Congress.[1] Another article on June 8 called into question the degree to which Congress was actually performing oversight on these programs.

The revelations were incredible for their insight into the misdeeds of the US security agencies. But they had additional significance in hacking circles, where commitments to both free speech and private communication are dearly held. On June 9 (during the unconference), the *Guardian* revealed the identity of the whistle-blower, Edward Snowden (with his permission).

1. Mirren Gidda, *Guardian*, "Edward Snowden and the NSA Files Timeline," August 21, 2013, https://www.theguardian.com/world/2013/jun/23/edward-snowden-nsa-files-timeline.

Stating in a video interview, "I have no intention of hiding who I am because I know I have done nothing wrong," Snowden claimed that the American people had the right to judge the NSA's programs themselves. On a lunch break that day, I was sitting with a Anika, a New York City–based woman around thirty years old, and, in what began as a casual chat, she told me that the whistle-blower had just "outed himself and is holed up in Hong Kong."[2] I knew what she was talking about enough to follow, but I was not up to date with this latest information, as I hadn't been checking the news much since the unconference had started, and the identity of the leaker was brand-new information. To my surprise, she then choked up and started to cry. Delicately, as we were barely acquainted, I asked if she was OK. She became more composed in short order and explained her profound connection with this stranger's actions. For Anika, Snowden's revelations as an NSA contractor who had blown the whistle on illegal government surveillance were not only of great importance in an ongoing debate about the correct balance between liberty and national security. She reacted in an immediate, visceral way to his personal sacrifice and identified greatly with his decision to leak this information to the press, fearing for his safety as he was now likely to be subjected to extradition and prosecution. Though she had little more information about Snowden than I did, as someone closely identified with hacking, what she perceived as his bravery and vulnerability hit her like a punch to the gut.

Hacking is not a single set of practices or cultural ethos. It is fruitfully understood by taking a genealogical approach, as suggested by sociologist Tim Jordan, which includes taking notice of what is absent in social phenomena as well as what is present.[3] This is particularly useful in understanding whether and how hacking has a relationship to politics and activism. Early threads of hacking, according to Jordan, were centered around altering technologies and manipulating information.[4] Hacking sensibilities also included the formation of communities, and the identification of coding as both a material practice and discursive resource. Free software played a special role, as programming "bifurcated between free software and the programming proletariat. The activity of coding is the same for both and individuals may, and often do, occupy both positions but there are distinct roles as a coder for community owned and collectively generated program[s]. . . .

2. Fieldnotes, June 9, 2013, San Francisco, CA.

3. Jordan 2016.

4. Jordan 2016: 5.

This is a broad distinction but it develops and begins to underpin many of the political aspects of coding."[5] Coding took on significance as both a material practice and a discursive resource, allowing people engaged in it to recognize themselves as members of a community and elevating both its status as a practice and the social status of its practitioners within that community. Though these developments may seem obvious from our present standpoint, they were contingent. Furthermore, their significance was not necessarily apparent in the moment, though it was at times crystallized through the circulation of texts like John Perry Barlow's "Declaration of Independence in Cyberspace" and hacker The Mentor's "The Conscience of a Hacker" (both of which are instantly recognizable as sacred texts that bind community members).[6]

Among hackers, there is *not*, in this genealogical telling, a commitment to political intervention, progressive or otherwise. In fact, hacker culture has displayed many politically regressive tendencies. As has been explored extensively, this includes a strong and exclusionary thread of masculinity.[7] In addition, some high-profile proponents of open culture such as Jimmy Wales, the founder of Wikipedia, have been heavily influenced by Ayn Rand and her so-called objectivist philosophy, which privileges individual power and entitlement.[8] David Golumbia has argued that *cyberlibertarianism* is a dominant mode of belief in Silicon Valley and represents a far-right libertarianism.[9] Lastly, the ethos of transgression favored by many hackers contains a strong potential for retrograde political effects.[10] This background is necessary to underscore that though this book has focused on the mostly progressive intentions of diversity advocates in open-technology cultures, the political waters they swim in have a history of being populated by people and ideas ambivalent or even hostile toward these intentions.

A later development in hacking practice involved hacktivists who were invested in building tools that were meant to secure the free and open (which

5. Jordan 2016: 6.

6. Jordan 2016: 7. Though Jordan does not spell this out, these future-facing titles also reach back toward Enlightenment tropes and values.

7. Jordan 2016; Jordan and Taylor 1998; Nafus 2012; Reagle 2013.

8. Rosenzweig 2006: 119. According to David Golumbia, "[Wales] is not merely an avowed libertarian but a follower of Ayn Rand so devout that he named his daughter after a character in Rand's novel *Anthem*" (2013b). Thanks to Nathan Ensmenger and Matt Wisnioski for help tracking down some of these traces.

9. Golumbia 2013a, 2013b; see also Borsook 2000; Kelty 2014; Turner 2006.

10. See Chen 2014; Coleman 2012b, 2015; Phillips 2015.

also meant secure) transfer of information over the internet.[11] This animated some strands of FLOSS with a somewhat different political valence. Since the emergence of the global justice movement (also called the *anti-corporate-globalization movement*) of the late 1990s, many technology-oriented activists have focused their energies on building tools that remake the internet with the goal of supporting activism.[12] This includes FLOSS tools to enhance privacy and anonymity in electronic communication, as well as activist-specific tools and software suites to permit activists to run email and other applications on independently built and maintained software (as well as hardware like servers), pushing back on state and corporate surveillance and enclosure.[13] Many scholars have noted that hacking can sound notes of revolt against commodification and alienated labor, more and less explicitly.[14] Lastly, leaking and whistle-blowing are also political acts that may be tied to a hacker ethos, as in the high-profile cases of Snowden and Julian Assange,[15] among others.

Scholarship on present-day hacking and making in the United States has asserted that much of the impulse that drives practitioners is not about politics. Rather, as Sarah Davies shows, much hacking and making is driven by a desire for community and for participation in leisure pursuit that one finds personally meaningful. This is related to claims that hacking can assume significance as unalienated labor, but Davies argues that a wider politics is actually circumscribed for many participants. She writes, "Contrary to expectations of the maker movement as heralding social change, the benefits of hacking were viewed as personal rather than political, economic, or social; similarly, democratization of technology was experienced as rather incidental to most hackers' and makers' experiences."[16] Earlier we have seen arguments about hacking versus making, with some practitioners claiming that making is more apolitical and more about lifestyle, while hacking is more politically charged. While this is a valid critique in some settings—hacking is

11. Jordan 2016: 10. Less important for the present discussion, hacktivists also launched electronic mass actions like civil disobedience and demonstration tactics in online settings (Jordan 2016: 10). Antecedents here include radical techies of the 1990s (Milan 2013).

12. Juris 2008; Milberry 2014. See Wolfson 2012 for a critique of this.

13. Milberry 2014.

14. Milberry 2014; Söderberg 2008. Though see Golumbia 2013a, 2013b for discussion of how some of these politics are potentially less progressive than proponents may intend.

15. Coleman 2017. Writing in the *New Yorker*, Raffi Khatchadourian stated that, "For Assange, there is no real difference between a hack and a leak; in both instances, individuals are taking risks to expose the secrets of institutions" (2017).

16. Davies 2017a: 1.

generally more politically valenced in Europe, for example—Davies's argument reveals a broader truth about the underspecified politics of hacking. This is an important point—for all the ink that has been spilled trying to nail down what the politics of hacking might be, it is definitely important to note that a significant proportion of contemporary hacking is essentially in line with the original iteration of DIY in the US, a project of suburban postwar homeowners' infusing a bit of "autonomous production" into their consumption practices.[17]

What can be concluded about hacker politics? In the main, to generalize about hacker politics is fraught. Anthropologist Gabriella Coleman has written of hackers' "political agnosticism,"[18] but over time she has chronicled an evolving politicization emerging as hackers have responded to perceived threats to their liberties and belief systems.[19] And yet, while it would be erroneous to conclude that hacker politics are monolithic, good or bad, neither are they neutral (to bastardize historian of technology Melvin Kranzberg's famous quote). Any formulation of *explicit* political stances occurs against the backdrop of general cultural assumptions, which have tended to emphasize altering and controlling technology, which has been related to a liberal—some would say libertarian—impulse that elevates the individual's exercise of freedom. Where this impulse leads is underdetermined, but not clearly in a progressive direction, nor necessarily tied to a notion of collective good. Part of what is so complicated here, according to Golumbia, is the plasticity that accompanies the ideological accretion around computing and digital technologies:

> Cyberlibertarianism's primary social and epistemic function is to yoke what would have previously been seen as at least liberal if not actually leftist political energies into the service of the political far right, with enough rhetorical padding to obscure at least partly, even to adherents, the entailments of their beliefs. In other words, cyberlibertarianism solicits anticapitalist (or at least anti-neoliberal) impulses and recruits them for capitalist purposes, to such a degree that many believers often do not notice and even disclaim these foundations.[20]

17. Gelber 1997. Gelber rightly points out how DIY was a practice of masculine identity construction, important when suburban homeowners might have had white-collar jobs removed from a hands-on, brawny masculinity (see Douglas 1987). While contemporary hacking expands DIY to include feminine craft/DIY practices, its practices are still largely (neo)traditionally gendered.

18. Coleman 2012a, conclusion.

19. Coleman 2017.

20. Golumbia 2013a: 3–4.

In other words, for Golumbia, even left-leaning political intentions have the potential to be taken up, transformed, or recuperated as part of a right-wing or even libertarian social and technical project, even in spite of people's intentions to the contrary.[21]

An updated genealogy of hacking would include feminist hacking. It would also potentially be broadened to include practices and practitioners who have been excluded by hegemonic accounts of hacking, as detailed in the historical introduction of this book. Feminist hacking is not necessarily immune from Golumbia's critique about the lurking potential for regressive tendencies, in part because feminism itself is contested and subject to corporate co-optation. But more importantly, for feminists and others, technology is a uniquely challenging site for progressive intervention.

The preceding discussion lays out hacking culture as a medium where diversity advocates' political intentions are being sown. The rest of this chapter follows the lines of diversity advocacy as they intersect with political stances that relate to diversity advocacy. As will be illustrated below, some political intentions expressed by diversity advocates are quite at odds with a right or libertarian political slant, which potentially inhibits their likelihood for growth in hacking culture. That said, their perhaps unlikely growth in this medium reflects both an urgent grasping for alternatives and the need to grapple intensely with the threat of recuperation by political agendas with which one does not intend to align.

A theme of this book is borders of care: reflexive attention to how boundaries are drawn around what counts as an object of care and intervention. How to set limits on what is a core concern (and thus meaningful intervention), versus what lies beyond, is a challenging aspect of diversity advocacy and a challenge to anyone analyzing this advocacy. As this book has argued, following Leo Marx, *technology* is a hazardous concept, in part because it enfolds many social and political issues; technology is politics by other means, whether or not this is explicitly acknowledged. Diversity advocacy, I have argued, uneasily blurs a line regarding whether it is intervening in technology for technology's sake, or whether it is addressing other matters. Though it is perhaps counterintuitive given diversity advocates' active, agitational stances—already chronicled in this book—I argue that their political

21. It is not coincidental that many believers do not necessarily notice or engage the contradictions. Rather, "negative liberty" is conceptually intertwined with "positive liberty," and a notion of collective freedom must include an account of individual freedom; they cannot be understood in isolation from one another, so potential for slippage in political valence is great (Berlin 1958 as discussed in Kelty 2014).

stances and intentions are not especially self-evident. In many cases, diversity advocacy sits alongside topics with overt political valences, hinting at them, flirting with them, or disavowing them. This chapter scrutinizes the ways in which diversity advocates navigate other political concerns, and then concludes with a meditation on what is gained and what is lost in technologists' isolation of the topic of diversity in the midst of a wider spectrum of political concerns.

The following three sections sketch some of the explicit political stances that are audible alongside diversity advocacy, including broader radical activism, antimilitarism, and anticolonialism. They are not mutually exclusive categories, of course. Crucially, I do not wish to imply that *all* advocates for diversity in open technology hold these political positions, merely that a nuanced accounting of the politics of diversity advocates reveals these stances to be salient in diversity conversations.

Radical Politics and Activism

A pastiche of political intentions was on display in the unconferences I observed. As described earlier, the organizing premise of the unconferences was to support women* in open technology, but this was a general principle. In practice, attendees themselves set the agenda; there was no prescribed program and people organized themselves into sessions based on interest. At the 2012 unconference in Washington, DC, someone proposed a session on "Intersectionality: feminism and other social justice work in technology circles" (figure 6.1). The session did not center on Kimberlé Crenshaw's use of intersectionality—intersecting social identities and their multidimensional, differential experiences of marginality and power—though this framing was in the background.[22] Rather, the session attendees discussed whether and how "geek feminism complements and works with other [social] movement stuff, or ways it might butt heads,"[23] in the words of one person. A second session on a related topic was "Using open stuff to create social change/feminist change."[24]

A common sentiment across these two days of sessions was that, in many attendees' primary open-technology circles, there was relatively little commitment to political engagement. (Quite obviously, those who

22. See also the 1977 Combahee River Collective Statement.

23. Fieldnotes, July 10, 2012, Washington, DC.

24. Fieldnotes, July 10, 2012, Washington, DC.

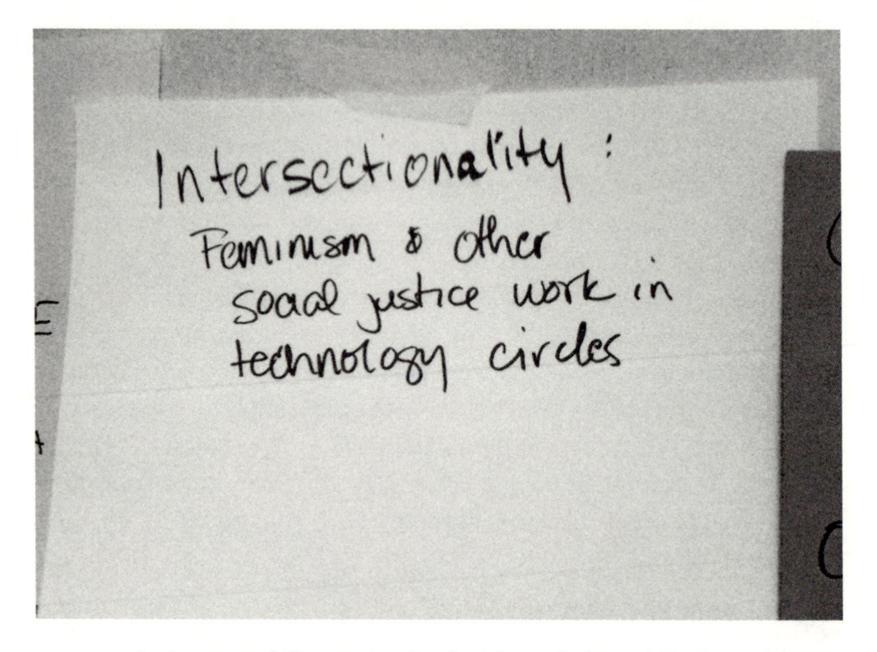

FIGURE 6.1. Session proposal, "Intersectionality: feminism and other social justice work in technology circles," Washington, DC, July 2012. Photograph by the author.

had self-selected to propose and attend these sessions were united in being politically engaged.) Many were puzzled at how far the world of open technology seemed from more engaged, activist stances. A woman from New York with a background in human rights advocacy asked the group, "Is it helpful to think of open source as a movement?" Another person replied, "Not necessarily. A lot of people in open source are not motivated by politics or social justice. It's actually pretty conservative."[25] Others agreed. Sam, a genderqueer developer from the San Francisco Bay Area reflected:

> There is a gulf between tech and [my background in] queer activism. "Tech for the common good" [as a framing] is missing an edge. [I think about] bringing a different or more radical kind of activism to open source or to tech. [If my workplace supports a cause like,] Let's make sure kids are web-literate, [I am thinking,] What about prison? There are people who *really* don't have access to communication.[26]

This quote is an excellent encapsulation of some of the dilemmas experienced by people who are oriented to social justice in an environment like open

25. Fieldnotes, July 10, 2012, Washington, DC.
26. Fieldnotes, July 10, 2012, Washington, DC.

technology. In Sam's telling, a respectable but anodyne workplace campaign to promote web literacy sparked her to consider communication rights in a more expansive manner, and to land on incarcerated people's communication rights (which are notoriously curtailed, not only through lack of access but through exorbitant rates charged by the telecommunications industry[27]). But she balked at raising this issue in her workplace, precisely because of its "edge"; incarcerated people's communication rights are a less well-known and potentially more controversial site for intervention, and might not seem like an appropriately middle-of-the-road, workplace-friendly campaign. In other words, though in her mind incarcerated people's communication rights and web literacy were related as matters of open technology and social conscience, this connection would not be apparent to everyone, even those invested in promoting open technology. It is of particular interest that she invoked her *personal* political commitments when thinking about the *workplace* campaign. As explored in the previous chapter, lines between workplaces and voluntaristic sites of association blur heavily in open technology. Yet there are established reasons for workplaces, even those committed to abstract notions of the common good, to steer clear of more radical political stances. Thus it is not necessarily surprising that an open-technology enthusiast who is also committed to political radicalism would bemoan missed opportunities for political intervention when an employer (for- or nonprofit) moves to support a more mainstream goal or campaign.

Sam also offered some illuminating commentary on the corporate clamor to recognize and support LGBT people in technology fields. (People in the session then kidded about informal community labels like *Homozilla* in the Mozilla community, *Glamazons* at Amazon, and *Oranges* at Apple, signaling a high degree of awareness of such communities in the industry more generally.) She said that on the one hand, she appreciated that large corporate workplaces in tech had acknowledged the presence of LGBT folks. But at the same time, company sponsorship of a happy hour ("if you're gay, there's beers here," in her words) or a Pride parade float stopped well short of her own commitments in struggles for social and political equality. She said, for example, "My history of activism or queer identity isn't gonna be the same [as a corporation's]. Gay marriage is not my battle, my politics are nonassimilationist."[28] These statements reflect tensions for queer communities that have been discussed at length in other contexts, and it is not my

27. Kukorowski 2012.

28. Fieldnotes, July 11, 2012, Washington, DC.

intention to rehash them here.[29] The main point is that many of the attendees of this unconference were not quite sure what open-technology fields and social justice work had to do with one another, even though some of them were fervently committed to both. Sam also remarked, "I have a background in queer activism, for twenty years. Then when I've come into the open-source community, it seems like there's a gulf. [Even the conversations that *are* happening] are Feminism 101. Is this what our community is about? Is this what are conversations are about?"[30] She wondered, "Who's even meant when we talk about 'our community'? For example, the Python community: two people who work in Java shops but [program in] Python, is that a community? People of color, queer people, [oppressed people], are who I mean when I say community—not SeaMonkey [an internet suite developed by the Mozilla community, which grew out of Netscape source code] users!"[31] Sam is white, but said, "I learned about race even though I'm white, I learned about disability even though I'm able-bodied, because this was the kind of activism I came up in."[32] Overall, her commentary indicates that she seeks to conjoin radical politics to open technology, but views her comrades in open technology as less politicized, or less politically sophisticated.

People repeatedly circled back to the topic of how they felt open technology embodied inchoate politics. One person said, "free software was bearded guys in the 70s that were supposed to be the radicals,"[33] but their legacy was less than she might have hoped. Other people speculated that community building and political consciousness building were particularly difficult in online environments. Sam, the queer developer from the Bay Area, said that she "would like to know more tactics for the internet because [it] is such a special, special place . . . a horrible place."[34] Here she is referring to multiple forms of antisocial online behavior for which online association is notorious, from flame wars to trolling, much of which is racialized, gendered, or otherwise intolerant. A woman from Boston who had extensive experience at the intersection of nonprofits and FLOSS said in response, "The tactic for the internet is to keep having one-on-one conversations over and over,"[35]

29. For example, Warner 1999.

30. Fieldnotes, July 10, 2012, Washington, DC.

31. Fieldnotes, July 11, 2012, Washington, DC.

32. Fieldnotes, July 11, 2012, Washington, DC.

33. Fieldnotes, July 11, 2012, Washington, DC. See Ensmenger 2015.

34. Fieldnotes, July 10, 2012, Washington, DC.

35. Fieldnotes, July 10, 2012, Washington, DC. She also said that she believed that non-profits were positioned well to understand and utilize the "freedom and empowerment" aspects

indicating that she thought people were more likely to be more reasonable with each other if they were not on display, and that free speech absolutism could be curbed when its potentially negative effects on others were less abstract; in other words, if someone objected to perceived hostile speech, the community would benefit from one person addressing another privately, sheltered from escalatory dynamics of piling on or shaming. A developer from Seattle agreed and said that in a project community she helped lead, they had "changed [poor dynamics] by making people see each other in person, but it took three years to turn around nasty, trollish dynamics."[36] All agreed that online congregation en masse coupled with the potential for anonymity—or even just the lack of face-to-face accountability—raised the possibility that voluntaristic groups would experience significant challenges having to do with governance and interpersonal or intra-group dynamics. While no one suggested that in-person groups were immune to negative dynamics, they agreed that groups whose main mode of association was online could be especially susceptible to negative dynamics, or even just the outsize impact of one or two "difficult personalities" or people with a high tolerance for adversarial interaction.[37] Political activism in open-technology cultures was understood to face unique challenges.

In another session at the unconference, people touched on their hesitancy to strongly politicize diversity advocacy itself. One person opined that she felt that it was unfortunate that she ought not use a "moral framing" because this might make people defensive about having been exclusionary, so she stuck to pragmatic arguments about building more widely used products and saying, "we're just not doing as well as we could be."[38] Liane, an organizer of the women*-only unconference said in an interview that she had raised money for the unconference by appealing to men in industry and saying, "wouldn't you like a more level playing field for your daughter?" but without dwelling on some of her own personal beliefs about power in these domains. She said, "[Many] men are playing a zero-sum game: in order to have wealth and power, you have to keep others from attaining it. There is a class-[and race-based] agreement, let's keep power with the white men."[39] In essence,

---

of nonproprietary software, but that "then you [FLOSS] called us [nonprofits] stupid [for not adopting FLOSS]," which set back relations. See McInerney 2009 for more on the nonprofit sector and use of FLOSS.

36. Fieldnotes, July 10, 2012, Washington, DC.

37. Reagle 2013; Reagle 2017.

38. Fieldnotes, July 10, 2012, Washington, DC.

39. Interview, July 24, 2014, San Francisco, CA.

she knew that expressing these beliefs to people who did not already share them would be perceived as confrontation, which would undercut her ability to fundraise. Along similar lines, Paul, who ran a number of programs to increase women's participation in Python, said that he often called his efforts *outreach* because this was unlikely to be controversial, whereas *diversity* as a label could set back his agenda, leading to wasting time trying to convince people that rates of participation in FLOSS differed according to one's social group, and that this differential mattered for the wider FLOSS community (what Sam had called "Feminism 101" conversations). And I observed a presentation on FLOSS "for shy people,"[40] which was another way to gently initiate a conversation about being more inclusive without touching the third rail and setting off a full-blown debate about autonomy and free speech, which many open-technology communities are poised to have at any moment. "Shyness" may have been a noncontroversial way into the topic of project cultures and interactions, but it tiptoes around the fact that unequal standing and unequal treatment have been experienced by open-technology participants. It deliberately avoids confronting the dynamics patterned on race, class, or gender that many people report as being a barrier to entry to participation in open technology.

Clara astutely drew several of these threads together. She had a capacious understanding of this terrain, having worked on various technical projects in both paid and volunteer capacities for many years, having quit a job as a systems administrator for a nonprofit institution to work as an advocate for open technology at a think tank, and holding personal political leanings that ran toward social and environmental justice and appropriate technology projects in Global South countries, especially in Latin America. She said that she assumed that some activists were deliberately framing their interventions in language that would enhance their chances of receiving funding but tended to obscure the extent of their radicalism: "[Diversity] advocates often have social justice motivations but work in for-profit organizations that aren't equipped, much less motivated, to do the right thing for its own sake. So the advocates learn the language that will get an initiative funded, or whatever their goal is."[41] In other words, diversity advocacy could give cover for more radical political intentions, but at the same time, these intentions could be diluted when framed as diversity projects instead of, say, incarcerated people's telecommunications rights, or queer antiassimilationist

40. Fieldnotes, June 2011, New York, NY.

41. Personal correspondence,—to author, December 28, 2016.

politics. The above comments reveal that for some people invested in diversity in open technology conversations, there is a self-evident (though sometimes rather nascent) connection between their engagement with technology and their wider political commitments. But as these comments expose, these aspirations sit together somewhat uneasily: technologists who envision deep connections between social justice activism and open technology are a minority in tech communities. "Diversity in tech" as a discourse may in some ways bridge these commitments, but it also bridges to agendas where a robust commitment to social justice is plainly more muted.

Opposing Militarism

At the 2012 Hackers on Planet Earth (HOPE) conference, a panel discussed "DARPA funding for hackers, hackerspaces, and education: A good thing?" in front of an overflowing audience. The panel was moderated by Mitch Altman, who had caused a significant ripple of controversy when he announced his refusal to participate in that year's Maker Faire, on the grounds that the Maker Faire organization had accepted Defense Advanced Research Projects Agency (DARPA, an agency of the US Department of Defense, DoD) funding. Featuring fire-breathing robots, product demos, and spaces for children to play with circuit boards, Maker Faire is an exposition space sponsored by O'Reilly media's *Make* magazine. It was first convened in 2006 in the San Francisco Bay Area and has since branched out to several more cities in the United States and Europe. It is a significant event and ritual space in the contemporary maker cultural movement.[42]

Known for inventing a "TV-B-Gone" remote control device that turns off televisions and for cofounding the high-profile San Francisco hackerspace Noisebridge, among other accomplishments, Altman is a widely respected figure in hacker circles and a regular feature at hacker conferences and in hacker media. The panel discussion included Altman; a hacker who had somewhat ambivalently accepted DARPA funding for a program in his hackerspace focused on space exploration and colonization; a hackerspace founder who posited that she had no ethical problem interfacing with the military in order to aid humanitarian intervention and thought hackerspaces were good places to discuss resiliency and "the coming zombie apocalypse;"[43] and a librarian involved in bringing hackerspaces and libraries

42. See Sivek 2011.

43. Fieldnotes, July 15, 2012, New York, NY.

closer together. They held a lively discussion, reaching consensus that this was an important topic for hackers to consider, but no further conclusion. As one panelist said, "One of the things I'm concerned about is, if you grow a piece of celery in red water, it's going to be red. If you are at Google, and you get 30 percent of your time to work on [your own] project, that's [still] going to be related to Google['s priorities], [so] I'm just wondering how this DARPA defense contracts money is going to influence these projects."[44] Another panelist said that a potential negative aspect of accepting this funding is that DARPA ultimately has a "military goal," which "means that anything you do in relation to them is, in effect, furthering the military prowess of the United States, which means you are in effect a secondary supporter of a machine that [exists] for the functional purpose of military conflict, in short, killing people or defending people." He drew a direct line between military conflict and DARPA funding, however distant this connection might seem from inside a hackerspace.

The librarian said that he worried that having DoD-funded hackerspaces in high schools would draw closer together STEM education and military recruitment. He said that "while DARPA does not recruit, military recruiters have a presence in high schools and they are extremely interested in STEM education."[45] He cited a recent article on military recruitment that encouraged recruiters to go into STEM classrooms, as well as research that indicated the prevalence of military recruiters targeting students from working-class backgrounds and communities of color. He was therefore concerned that hackerspaces in high schools might be used to facilitate military recruitment, which would be illegal (because under US law it is illegal to recruit children for military service who are under seventeen [and under eighteen under international treaty[46]]) and, further, unjust (targeting youth from poorer and racially minoritized backgrounds for military service perpetuates these communities shouldering a disproportionate burden).

It is important to note that 2012 was only a few years into the global economic crisis caused by the financial industry's reckless gambling in the subprime mortgage market in the United States. The moment at which DARPA was offering grants to hackerspaces and to Maker Faire (to put makerspaces in high schools and thus cultivate STEM mentoring and education) was also

44. Fieldnotes, July 15, 2012, New York, NY.
45. Fieldnotes, July 15, 2012, New York, NY.
46. Child Soldiers International n.d.

a moment that saw the slashing of the budgets of public schools and libraries, often quite drastically. The librarian on the panel felt that grant-based (and especially DARPA) funding clashed with the civic mission of libraries and the ethos of most hackerspaces. He said:

> There are very few neutral places left in the world. And if you think about the public library, anyone can come to the library and read basically whatever the hell they want, look up whatever you want. We are completely neutral, and we'll defend your right to read and look at whatever you want until the bitter end. That's why we're here. To a large extent, that was my original interest in hackerspaces. The model is very, very [similar.] It's a neutral space where you can come in, you can destroy whatever the hell you want, set it on fire, we don't care.[47]

By contrast, he felt that schools and hackerspaces would necessarily be affected by having DoD sponsorship and would no longer support more "neutral" forms of exploration or knowledge creation.[48] He also expressed concern that the DARPA grants to fund makerspaces in high schools would be unevenly administered, with wealthy high schools having more ability to control how their makerspaces were run, and other schools having fewer resources and less quality control: "We're seeing the [funding for public institutions] go like that [hand gesture of decline]. It's gonna be, like, substitute teachers working for a couple dollars a day, with instructions on a sheet of paper [that say] 'you just teach what's on this paper and shut your face,' and then the military recruiters [step in and] say, 'Yeah, I can do that for you.'"[49] In other words, he was concerned that DARPA involvement in education might both reflect and contribute to economic and social stratification; he saw an alarming pattern in civic and public institutions losing funding and militarized institutions stepping into the breach. He also discussed the possibility that students would have less freedom at DARPA-funded spaces, with less making (or destroying) "whatever the hell you want" and more focus on designs for artifacts like—he posited—unmanned aerial vehicles (UAVs, drones). He cited literature that, he claimed, could be read as a statement that a goal for DARPA in these educational spaces was to "inculcate

47. Fieldnotes, July 15, 2012, New York, NY.

48. Of course, knowledge is never neutral. Yet his point about the sudden loss of municipal funding for civic, public institutions stands; see Klein 2007.

49. Fieldnotes, July 15, 2012, New York, NY.

FIGURE 6.2. A "friendly surveillance drone" addressing children, satirized by cartoonist Tom Tomorrow, 2012. Copyright Tom Tomorrow, https://thismodernworld.com/.

a military design methodology in high school kids"[50] (figure 6.2). Other panelists pushed back on this, saying that DARPA was unlikely to recruit for the armed services directly, or even for DARPA itself, because DARPA usually contracted with industry for engineering work.

There is merit in the latter claims; in point of fact, DARPA is not directly involved in military recruiting, and supports much military research and development in industry through funding and contracts. Yet in a wider sense, the librarian's concerns are still pertinent. While it is not necessarily realistic for advances in applied military UAV technology to come out of a public high school, it is certainly the case that the technological designs and applications that are valorized in an educational setting may affect how students view technology and their own relationships with STEM artifacts and careers. And is not a stretch to imagine that such a setting would be a

50. Fieldnotes, July 15, 2012, New York, NY.

space for ideating an engineering career; indeed, this is a goal shared by educational institutions, Maker Faire, and DARPA alike.[51] To have much of the play and education in a makerspace center on, for example, radio-controlled flying objects, and to deny that there is a potential relationship between these objects and militarism, is to be disingenuous at best, considering the history of flight technologies, radio communication, and electronics miniaturization.[52] And in subsequent years, ties between makerspaces and military contractors have increased: as of 2017, Northrop Grumman Aerospace Systems runs at least five makerspaces.

Moving beyond the more abstract panel discussion, some of these issues can be observed in situ at the Philadelphia hackerspace whose membership had forked in the aftermath of conflict over a sex-toy-hacking workshop (described in chapter 3). Clara, the former board member who had left the hackerspace, later told me she thought militarism was another unstated aspect of the conflict over values in the space. Though no longer active in HackMake, she continued to occasionally read traffic on its email list. She described a posting where one member of the hackerspace had enthused on the space's electronic mailing list about a design for a one-winged UAV designed for surveillance. Similar to the shape and properties of a maple tree seed spiraling through the air, a prototype had recently been tested and a video released of its flight (figures 6.3 and 6.4).

Hackerspace member Ian posted the video on the mailing list, along with his congratulations to Jim, another member: "Look what Jim's been doing! Good Job Jim!" Since Clara knew all the players, she explained that Jim had, at work, been involved in the development of the prototype, which Ian, Jim's friend and cohobbyist, thought was neat and wanted to share with other space members.[53] She said, "[Jim] is a very, very sweet and smart member who happens to work for Lockheed, I think. but [military-industrial]

51. See O'Leary 2012. Drones in particular have been taken up with remarkable fervor by the DIY/*Wired* magazine set. In 2009, *Wired* magazine editor Chris Anderson launched a new venture, a website entitled "DIY Drones," which by 2012 boasted over 30,000 registered users. He described the mission of DIY Drones as "open sourcing the military industrial complex. . . . In two years, we have begun disrupting a multimillion-dollar industry with the open-source model. . . . We can deliver 90 percent of the performance of military drones at 1 percent of the price" (Takahashi 2012).

52. Thanks to Peter Asaro, Chris Csíkszentmihályi, Seda Gürses, Wazhmah Osman, and Madiha Tahir for conversations about these topics; and see Asaro 2012; Osman 2017; Parks 2016.

53. The prototype received celebratory treatment in tech press, such as a 2011 article on Geek.com entitled "Lockheed Martin wants a fingertip-sized UAV in every soldier's backpack" (Humphries 2011).

FIGURE 6.3. UAV prototype modeled on a maple seed, 2011. Courtesy of Lockheed Martin Corporation.

FIGURE 6.4. Maple seeds. Courtesy nlitement/Wikipedia, CC BY-SA 3.0.

complex for sure."[54] In her time with the hackerspace, she claimed that no one had ever explicitly questioned whether their hobby shaded into technology that was being sponsored by the military or developed by military contractors like Lockheed Martin. But she didn't think this was because no one was thinking about it; rather, she said that "[because] of social ties, there's a politeness incentive to de-politicize the [conversation] . . . [But that] larger topic [does come] up a lot at my dinner table b/c we have 3 MIT grads under our roof[.] Where do MIT grads go to work? Sili[con] valley or defense contractors . . . There's actually a lot of social incentive to be polite about the politics of this stuff."[55] She noted that on the mailing list, another member, Alan, next proceeded to ask Ian about the technical aspects of the prototype. She astutely pointed out that Alan may have diverted the conversation about the prototype to a discussion of its physical properties as a strategy to avoid what might be an uncomfortable conversation for members about politics, which no one wanted to have. (This email exchange occurred in 2013, after Altman brought the topic to the fore in hacker communities in 2012.)

A few years later, in an interview, another member of the hackerspace, Liam, told me that the space had mostly avoided an overt discussion about militarism, in part by passing up opportunities for grant funding that might bring the issue to a head among members. He said the question was moot because "we're not cool enough for DARPA funding. We've got a member who works for the navy, he does cool top secret shit on his day job, [but] he comes to us [to make an] art car for a sculpture derby."[56] In this statement, he implies that "art car for a sculpture derby" is both too low-tech and too far afield to raise any issues of the presence of militarism in the space. He also suggests, likely accurately, that this member of the space is simply a technology enthusiast who enjoys tinkering and making in his spare time.

Then, somewhat to my surprise, Liam also volunteered that a member who had owned a lot of guns had passed away. He said, "He was the one who would bring guns to the space . . . He [was] a friend, you'd say 'I don't feel comfortable [with that],' but he doesn't care . . . [Also] in the case of the zombie apocalypse, he's the one I'd go to." This comment suggests that his own discomfort with the other member bringing guns to the space was mitigated by their friendship and was not great enough for him to draw sustained attention to the matter. Though passed off in a rather tongue-in-cheek, silly

54. Shared in personal electronic chat with author, April 11, 2013. Names are pseudonyms.

55. Personal electronic chat with author, April 11, 2013.

56. Interview, July 29, 2016.

comment about the "zombie apocalypse," his comment and a similar one above by a HOPE panelist does suggest that hackerspace members are not unfamiliar with survivalist tropes. I also observed this in the Oakland makerspace led by people of color. Notably, the Oakland makerspace's use for weapons needed after the apocalypse included hunting and archery, but not guns explicitly. They were perhaps imagining a gentler and less well-armed future than some other hackers.

In any event, given that Philadelphia has a strong pacifist Quaker tradition, and that hacker culture had begun to debate whether it should accept DARPA funding in 2012, it seems likely that by the time of this 2016 interview HackMake had a culture that tacitly endorsed militarism enough that people who would be especially troubled by weapons in the space or by uncritical celebration of military-contractor-developed technology would probably join other local spaces instead (perhaps the one to which the would-be sex toy hackers decamped). Liam concluded, "We avoid the ethical discussion by not seeking funding, we don't do grants. We'd have people on both sides of that debate, for sure."[57] It is interesting that he suggested that their membership structure, based on dues and not grants, allowed them to skirt ever making this an open topic for debate or forcing the hackerspace to codify its position. In any event, this also bolsters Clara's observation that people often avoided bringing up these topics out of a "strong social incentive to be polite." Given that this was a technical hobbyist space, not a politically oriented one, and that members were known to work for both the military and companies that were recipients of military contracts, it is understandable that to bring up one's concerns about militarism might seem confrontational. But the level of comfort with weapons and military technology some members exhibited was not something she shared. Indeed, even Liam, who was a close friend of the person who brought guns to the hackerspace did not feel it was especially appropriate, but neither did he make an issue of it.

In spite of HackMake's tacit acceptance of the presence of militarism in their space, it is clear that some in the hacking milieu were less than acquiescent. Altman, for example, was opposed to the normalization of militarism in hobbyist spaces. Another panelist at the 2012 HOPE conference (on a separate panel) alluded to these issues in his talk when he quipped, "I think it's great that DARPA is now working with *Make* magazine because the next inventor of napalm could be a kid."[58] This barbed, sarcastic comment stood

57. Interview, July 29, 2016.

58. Fieldnotes, July 15, 2012, New York, NY.

in contrast to other stances hackers might adopt about weapons technology; an "apolitical" technophilia might incline some to argue that weapons are "neutral tools" to be figured out and mastered.[59] On the panel about hackerspaces and DARPA funding, another panelist argued that, even though she "might wish" the military were not sharing cultural space with hackers, it was naive to not accept that they were here and try to learn what could be learned about technology and/or use their presence to promote technological enthusiasm. Specifically, speaking of drones, she said that a "quadcopter" (UAV design) had "delivered [her] a beer the other day, and that was pretty cool."[60] (This is an instantiation of the "neutral tool" argument, of course.)

It is to be expected that people expressed a multiplicity of ideas and values with regard to militarism in hacking communities. Altman launched the panel discussion by stating that the hacker community was one he valued for its range of opinions, and what he primarily hoped to accomplish was a respectful discussion that would make people think, which he said was what he most embraced in this community. Being thoughtful, and even being wary of weapons technologies, does not necessarily dispose one to opposing militarism. Sometimes people who have extremely strong feelings about weapons technologies, like atomics weapons scientists, hope to engage with these technologies in order to control them (and even prevent their use). Other reasonable people might disagree and choose to engage in activism around disarmament.[61] That said, hacking communities demonstrate in some cases a lack of consideration about whether their civilian activities potentially amount to military public relations (or even research and development). For reasons including technophilia, belief that technologies are neutral tools, and not wishing to have uncomfortable political conversations in hobbyist spaces, hackers often appear to sidestep the potential politics of the artifacts with which they are joyfully tinkering. Some diversity advocates are more attuned to these thorny ethical issues. One person in the 2012 unconference for women* in open technology noted in the session on soft circuits (sewing with conductive thread to make electronic circuits, described in chapter 4) that these creations could have military applications, even though they seemed like domestic crafts projects. And as Clara put it, "These folks [mainstream hackers] are generally not disciplined appropriate technologists, so they're not aware of, much less [trying to build] around,

59. Winner 1988.

60. Fieldnotes, July 15, 2012, New York, NY.

61. Gusterson 1996.

the issues of communities who are victims of all the militarized stuff to begin with."[62] For Clara, there was a difference between being a techno-enthusiast and someone who had thought deeply about power relations and ethics in relation to technology—in other words, a "disciplined" technologist.

Some of these issues gained wider traction in 2017 and 2018, the first years of the Trump administration in the United States, when organized coalitions of tech workers challenged their firms on major government contracts. Two prominent examples were workers advocating that Microsoft not build a database that could assist Immigration and Customs Enforcement (ICE),[63] and that Google cancel its involvement in Project Maven, a proposed deployment of algorithmic machine learning in war zones.[64] Obviously, there is a long history of interdependence and collaboration among industry and the military and academia,[65] so it remains to be seen to what extent these challenges may be effective in articulating less militarized visions of tech work.

Global Positioning Systems: "Decolonizing Technology"

Clara's comment about "communities who are victimized by militarized stuff" raises the topic under consideration in this final section of this chapter.[66] In interviews, field sites, and online, some advocates for diversity in open technology pondered not only identity categories (further explored in chapter 7) but also reflected on how difference is generated through spatial, economic, and political relations. Such advocates expressed concern that the technologies they on the one hand embraced so fervently were on the other hand implicated in colonialist or neocolonialist relationships, which undercut their liberatory potential.[67] Diversity advocates sought to name and explicate these relationships as an antecedent to recuperation of technology.[68] Self-identifying as a "white-passing Latinx" person who had lived in and worked

62. Personal electronic chat with author, April 11, 2013. See Pursell 1993 on appropriate technology.

63. Upadhya and Tech Workers Coalition 2018.

64. Tarnoff 2018.

65. Edwards 1996.

66. Chakravartty and Mills write, "Information and communication technologies (ICTs) are inseparable from the emergence and expansion of modern capitalism, colonialism, and racial regimes of militarized power, yet they have remained peripheral to decolonial critique" (2018: 1).

67. Irani et al. 2010; Philip et al. 2012.

68. See Toupin 2016b for an explication of anticolonialist hacking in 1980s South Africa.

on projects in the US, Europe, and Latin America, Clara herself had a highly reflexive awareness of relational differences in power and position.[69]

For example, in their call for participation in the 2016 feminist hacking convergence, organizers wrote,

> We take as a starting point the assumption that colonialism has invaded and embedded the digital realm and our technologies in general. . . .
>
> How then can we imagine the decolonization of technologies and of cyberspace? What would such processes, epistemologies, and practices entail? How can feminist anti-colonial, post-colonial, and/or indigenous frameworks shape and strengthen our analysis in our collective reflection on such questions? . . . We will explore the intricacies of colonial technologies while at the same time trying to conceive what decolonial technologies mean.[70]

Here the organizers invoke feminism, indigenous,[71] and postcolonial or anti-colonial frameworks and suggest they be brought to bear on technologies and cyberspace, suggesting that the best possible basis for praxis requires analysis and reflection on these issues. In practice, this orientation was rather vague. As postcolonial theorist Edward Said writes,

> "The colonized" was not a historical group that had won national sovereignty and therefore disbanded. . . . The experience of being colonized therefore signified a great deal to regions and peoples of the world whose experience as *dependents, subalterns, and subjects of the West* did not end—to paraphrase from Fanon—when the last white policemen left and the last European flag came down. To have been colonized was a fate with lasting, indeed grotesquely unfair effects. . . .
>
> "The colonized" has since expanded considerably to include women, subjugated and oppressed classes, national minorities. . . . The status of colonized people has been fixed in zones of dependency and peripherality, stigmatized in the designation of underdeveloped, less-developed,

69. Clara said, "Being in [Philadelphia] is an education in how white I am. Being in Germany was an education in how brown I am. And being in El Salvador, I'm American and also family/La Raza" (personal correspondence with author, June 9, 2019).

70. [Feminist Hacking Call for Participation] 2016.

71. See TallBear 2015 on "twenty-first-century indigenous knowledge articulation" with regard to human/nonhuman boundaries, which implicates science and technology as they are currently constituted; Vowel 2016 (chapter 8).

developing states. . . . To be one of the colonized is to potentially be one of a great many different, but inferior things, in many different places, at many different times.[72]

In spite of rigorous empirical and theoretical attention to these issues in certain academic subfields, as Said's quote illustrates, there is a reason for the looseness surrounding discussions that invoked colonialism (or decolonization) in open-technology circles. Even as a nonspecific concept, it satisfied advocates' desire to attend to geopolitical, spatial, and economic inequities deriving from the hegemonic positions of certain nations or peoples relative to others. In terms of advocates' use of these concepts, we might take Said's description of "the colonized" as "dependents, subalterns, and subjects of the West" as both literal description and figurative approximation (and indeed Said goes on to describe how this concept has been malleable enough to include various "subjugated and oppressed classes" such as women, national minorities, and others[73]). In a variety of forums, Global North technologists reflected on the interplay between their encounters with technologies (especially their commitment to hacking, and their association of technology with agency) and the troubling and polluting origins of those technologies, as well as technology's ongoing implication in Western economic and political dominance, including settler colonialism.[74]

A variety of empirical examples peek through field sites. Soliciting participation in the feminist hacking convergence, organizers recognized that their event would be convened on colonized land. In their words, "We acknowledge that we are proposing the [event] to take place on the traditional territory of the Kanien'kehá:ka. The Kanien'kehá:ka are the keepers of the Eastern Door of the Haudenosaunee Confederacy. The island of 'Montréal' is known as Tio'tia:ke in the language of the Kanien'kehá:ka. Historically, this location was a meeting place for other Indigenous nations, including the Algonquin peoples."[75]

During the event, participants acknowledged that their acts of cultural and material production were situated within the Global North and dependent on materials and labor extracted from the Global South.[76] As described

72. Said 1989: 206–7. Emphasis added.

73. Said 1989: 207.

74. Aouragh and Chakravartty 2016.

75. [Feminist Hacking Call for Participation] 2016.

76. See Ceruzzi 2003: 193; Eubanks notes that even within the United States, a "high-tech information economy" is propped up by low-status workers (2012).

in chapter 4, a burgeoning Montréal activist hardware group hoped to assert autonomy from labor, extraction, and postuse processes they perceived as exploitive by repairing discarded computers. They sought to shorten and localize the chain of production and consumption, recirculating repaired machines running free software within their local community. There is reason to believe that this would be an extremely difficult process to enact on a large scale; computer hardware is, quite simply, not designed for long life or simple repair. It is, conversely, designed for obsolescence: "a high rate of machine turnover marks a condition of tremendous profitability for the computer hardware and software industry[ies]," in the words of Jonathan Sterne.[77] Nonetheless, a member of this aspiring group listed as their goal, "Recycling with partners that don't ship to dumps overseas, getting to zero waste, or as close as possible."[78] (North American electronics recycled by well-meaning consumers and institutions often wind up in unregulated sites in the Global South such as Agbogbloshie, outside of Accra, Ghana, where precious metals are reclaimed under conditions that are extremely hazardous to workers' health.) Fervent about hacking, the aspiring Montréal collective sought to reposition it as a social and material practice. Aware of the dirty and exploitive conditions of electronics' production and postuse, they hoped to divorce hacking from the exploitation of people in the Global South and the environmental degradation upon which it currently rests.

A promotional poster for a feminist hacking event convened at an anarchist autonomous collective in Catalonia by some of the same people a couple of years earlier suggests similar critical awareness of Global South–Global North relations. Here, feminist hackers have appropriated an image (*Uterine Catarrh*) by Wangechi Mutu, a Kenyan artist working in the United States, and imposed on it text advertising their event, a "trans futuristic cyborg cabaret," including the words "gender hacking, anti capitalism, libre [free] culture, anti-racist, anti-sexist" (figure 6.5).

Mutu's piece depicts a gynecological exam (replete with scientific instrument—the speculum—and flowing white vaginal mucus, "catarrh") set between the eyes on a white skin mask partially covering the face of an African woman depicted in a hyperprimitivist fashion. There is a vivid grotesquerie to this visage, which absorbs a colonial and scientific gaze, and maybe gazes back. It is impossible not to wonder if there is a deliberate association with Frantz Fanon's famous book exploring colonial domination,

77. Sterne 2007: 20.

78. Fieldnotes, August 16, 2016, Montréal, Canada.

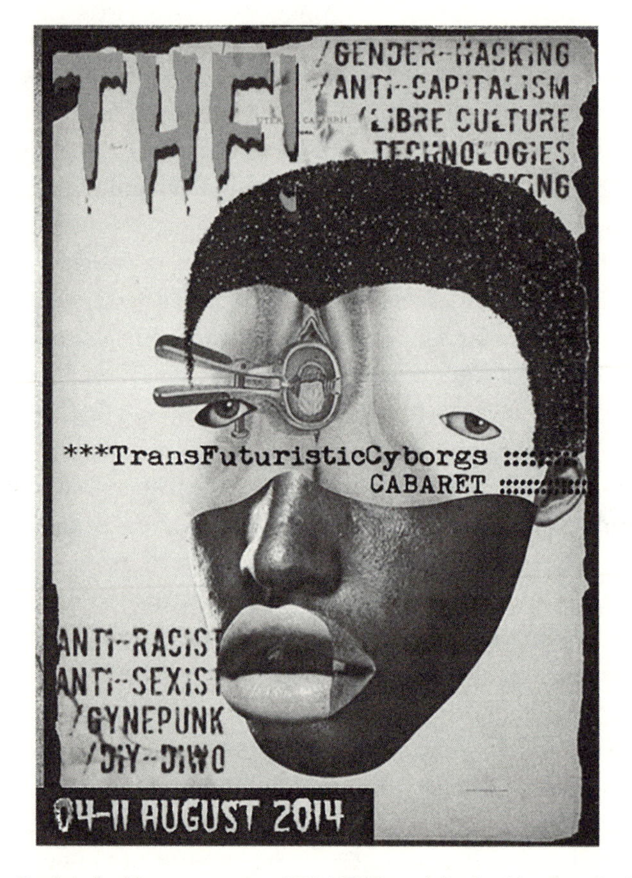

FIGURE 6.5. Feminist hacking event poster, 2014. DIWO, or doing it with others, is a less heroically self-reliant DIY.

Black Skin, White Masks. Thus, a politics of liberation including anti-racism, gender hacking (echoing cyberfeminism), and free culture are all linked for the feminist hackers.

Other actors talked about these dynamics in less abstract terms. Perez, a US-based activist who identifies as a "bi-cultural Latina" said in an interview, "Some of my closest friends are developers from Cuba, they didn't learn how to *really* code [code extremely well] until they moved to Spain [where computing resources were not scarce]. [In Cuba,] the Cubans had to build their own computers, find parts, construct their own computer. University gave you some hours on the internet, but speed is really slow. It's a testament to how smart they are to figure stuff out in these environments."[79]

79. Interview, Perez, November 4, 2016.

She described the Cubans initially learning programming by writing out code longhand with pencils at home, and then running it on university machines later, with much lag time in between. She compared this with her own experience learning different programming languages, where, she said, if she encountered a problem her existing knowledge base didn't prepare her for, she could just "go in and Google it. . . . If I didn't have the internet, I wouldn't have been able to do that." Perez observed that while her Cuban friends might feel their coding skills were inferior when they encountered people from the United States or Spain, she saw this in opposite terms: they had learned so much with scant resources, which demonstrated grit and ingenuity beyond that of many programmers who had learned their skills in the Global North.

Relatedly, STS scholar Anita Chan found that FLOSS activists in Peru were enthralled by the possibilities they imagined for upending historical dependencies experienced by their nation and region of the globe. However, they encountered a disconnect between their own interests in programming and their university computer science classes. She quotes an activist recalling a dean who had said he "should not dream of working with Bill Gates in Redmond," and was discouraged in his university study from learning more advanced programming associated with operating systems, as this was "not for us [Peruvian] students."[80] According to him, the university administrator reinforced a notion that innovation in computing was not to originate in Latin America; these countries were destined to be dependent on technology that would be created elsewhere. The activist perceived great limitations in the university-sanctioned vision of computing education, which urged him to orient to the needs of established local businesses.[81] By contrast, when the activist came to FLOSS, he ascribed to it meaning as a tool for agentic personal growth in the vein of hacking, and as one with national significance, offering "new potentials for developing other pathways towards [Peruvian] technological, political, and economic independence," in Chan's words.[82] Of course, using closed proprietary software such as Microsoft cost

80. Chan 2013: 160. See also Margolis 2010 and Sims 2017 for discussions of these dynamics in US education.

81. In her study of Indian coders working in Germany, anthropologist Sareeta Amrute notes that Indian coders were bound to a particular type of coding work: enough to get their foot in the door in terms of employment but limited to work that is shorter term and associated with lesser accomplishment (2014: 110). Thus, power relations circumscribe the kinds of bodies that are associated with high-status coding work in ways that correspond to economic and political power. See also Rodino-Colocino 2012.

82. Chan 2013: 141. See also Medina 2011.

Latin American governments and institutions a great deal in licensing fees, which FLOSS by definition did not. Technological and political sovereignty are bound up together in numerous ways.[83]

This being said, FLOSS still struggles in many ways with the hegemonic presence of Global North culture. In his ethnography of FLOSS in Brazil, Yuri Takhteyev describes how Brazilian FLOSS programmers predominantly code in English because it keeps their code consistent with global, professional coding practices—even though it is an option to use Portuguese for some aspects of their code—and because English is already encoded into the grammars of the programming languages themselves.[84] Others echoed the notion that FLOSS cultures themselves needed examination. On an email list discussing an upcoming unconference (that was never convened), one person said that she would be interested in talking about "free culture/open source from a decolonial/critical perspective that challenges the rather imperial ethos embodied by organizations [like] the Electronic Frontier Foundation (basically trying to approach the 'free' movement from an *anti-oppressive* place rather than the anarcho-libertarian ethics of most of the orgs run by white men)."[85] In other words, she claimed that mainstream or Western FLOSS culture had not examined how it potentially reproduced imperial power vis-à-vis groups trying to enact "open culture" in other places, with different emphases and values. Another woman, based in both the United States and Latin America, elaborated in a list discussion about white or Global North feminism, "[A]s someone that lives between US and latinoamerica I can totally see the two different FLOSS worlds. . . . The different organizers need to really understand . . . the importance of hav[ing] a diverse tech community (that also [extends beyond] gender diversity)."[86]

The Montréal convergence's intentions to explicitly attend to the topic of decolonizing technology were in practice undercut by the fact that many of the people they had hoped would attend the event to lead workshops on these issues failed to show. Some of us met to discuss the topic anyway. One

83. Couture and Toupin 2017.

84. Takhteyev 2012, chapter 2.

85. Email,—to [Women in Open Tech List], March 18, 2015. Emphasis added. This resembles Sandra Harding's critique of "value-free" science, which shifts the burden of science away from "neutrality" and toward a starting point that takes into consideration the needs of disenfranchised groups (2016).

86. Email,—to [Women in Open Tech List], July 31, 2015. Chandra Mohanty writes, "Western feminisms appropriate and 'colonize' the fundamental complexities and conflicts which characterize the women of different classes, religions, cultures, races, and castes." (1984: 335); see also Collins 2000; Combahee River Collective Statement 1977; Crenshaw 1991; Haraway 1991b; hooks 1989; Mohanty 2003; Sandoval 2000.

of the convergence organizers explained that the subject had originated at an event on Afrofuturism, where a panelist had raised the issue of cultural diversity in hackerspaces, not only gender diversity. She also said that she had reached out to people at *eco-camps*, antiracist, and indigenous-led groups she knew. It is impossible to know why the people she had solicited did not attend, and it is certainly true that this event—which took place on the heels of the World Social Forum (WSF) in Montréal—was competing with other meetings and events, and was weakened also by significant levels of activist fatigue after the WSF. Nonetheless it is notable that this event was a very genderqueer but mostly white radical feminist hacking meeting. The absent activists perhaps approached the intersection of technology and liberatory politics differently, possibly with critical ambivalence,[87] or they may have equivocated about attending a mostly white event.

Though many of the above examples are relatively utopian and even abstract, they suggest that some Global North activists who had been thinking of diversity in more identity-based terms were starting to think about geopolitical power relationships and how this affected the terms of their activism. Advocates were, of course, invested in a high-tech, digital future; they were, after all, brought together by a shared interest in FLOSS computing and hacking more generally. At the same time, they were committed to locating the artifacts and practices of open-technology production within culture, space, and history.

By contrast, some actions by industry players underscore the obduracy of relations critiqued by the above advocates. In 2015, Mark Zuckerberg pushed to introduce a service called Free Basics in India, which would have given people complimentary limited internet access, with Facebook at its center. In 2016, the Indian government banned Facebook's service, citing net neutrality.[88] Critics such as American technologist Anil Dash chided Zuckerberg, advising him to "reckon with the history of western corporate powers enforcing their desires on a broad swath of the Indian population, especially India's poorest . . . a colonialist 'trust us, it's for your own benefit' pitch is a hard sell."[89] Indeed, technologists are often implicated in impulses to solve perceived problems in the Global South with Global North tech fixes. At a Google-hosted reception for the 2013 women in open technology unconference in San Francisco, attendees were ushered into a hall where a video played, featuring women entrepreneurs in a hardly

87. Eubanks 2012.

88. Lyndgate 2016.

89. Quoted in Lyndgate 2016.

elaborated African country who produced handmade bracelets, underscoring Google's commitment to fostering women's value and ingenuity;[90] the video also celebrated practices of financialization through microlending.[91] Audience members were then encouraged to look under our chairs for gift bags, Oprah-style, which contained Google swag such as branded water bottles, lip balms, and the bangles featured in the video, which revealed themselves to be iterations of Google's logo (four thread-covered hoops in Google's colors; I wound up taking mine home but always felt I had no use for them and eventually donated them to an internal recycling area of my apartment building when I moved, a few years later).

This feature of the Google reception (like much of the rest of it) was met with dismay by the unconference attendees, many of whom were not convinced that this pablum-laced presentation had much to do with the gender-in-tech issues they perceived as important. The video imputed to its audience a feminine reform mentality and belief in commodified (branded) empowerment. While many in the room were women who cared about Africa in a broad sense, there was palpable discomfort about being drawn into these issues on the terms laid out by Google, which contained an undercurrent of "saving the Other." I inferred that many were also at a loss about what to do with feminine jewelry, even more than I was. With this audience, which was largely ambivalent about traditional femininity, and suspicious of white savior narratives, Google's bangles were an awkward misfire.[92]

Conclusions

As noted in the introduction of this chapter, it is difficult to generalize about hacker politics, but we have seen moves in explicitly political directions with the rise of hactivism, and hackers sounding notes of resistance to commodification and alienated labor. (This is not to say that "apolitical" strands of hacking ever actually lacked politics, of course.) What the above discussion demonstrates is that feminist hacking should be included in an updated genealogy of hacking.[93] Of course, feminist hacking, like feminism itself, is

90. Fieldnotes, June 7, 2013, San Francisco, CA.

91. See Poster and Salime 2002.

92. In a discussion of a "confidence you can carry!" charity handbag campaign (whose "confidence as a commodity" contours mirror some of the discourse at this event), feminist media studies scholar Sarah Banet-Weiser notes that traditional definitions of femininity are "overly focused on fashion, accessories, and the body" (2015: 185).

93. See Davies 2017b; Fox and Rosner 2015; Gajjala 1999; SSL Nagbot 2016; Toupin 2015, 2016a.

not monolithic, though it may be safely be generalized as fusing hacking with values of care. The chapter's opening vignette, wherein the unconference attendee was moved to tears by the actions of Edward Snowden, illustrates care: Anika burst into tears because of the intensity of her affective care for Snowden's well-being, and for the political ramifications of his whistleblowing. Far from *pwning* or dominance, hacking here was closer to platonic love, as elaborated by bell hooks: it included care, trust, honesty, and commitment (to community and to principle).[94] A couple of years later, I interviewed Anika more formally in New York, and she spoke of herself valuing what she called "the numinous."[95] She explained that one of her parents had training as a priest and she understood both coding and teaching as having almost spiritual dimensions; the joy and the intensity of them were in their best moments shot through with something almost divine.

This chapter has broken out a few strands within feminist hacking—radical politics in general, antimilitarism, and anticolonialism—from the concerns of diversity advocates. In describing these topics, I in no way wish to imply that advocates have necessarily accomplished their intentions; I outline them in order to demonstrate that these topics have attracted attention and given rise to articulations. These topics are important because they shed light on what is at stake for the people using hacking to critique both hegemonic hacking culture and aspects of the world at large. Diversity is but one of many political goals these hackers of community are seeking to effect in open-technology cultures.

Advocacy for diversity in tech may seem like a potent issue of justice. This is especially true in our current moment, in which men's rights activists challenge women's presence in the electronic public sphere,[96] when neo-Nazis coordinate online to march emboldened in the streets, and when even Google firing engineer James Damore over workplace expression of his views on gender essentialism is a flash point. But it is important to remember, as Golumbia cautions, that some left or liberal impulses in tech have the potential to be recuperated for purposes that do not necessarily align with their proponents' intentions. Diversity, though a wedge issue in a culture war, is in fact having a long moment where it is becoming part and parcel of corporate and neoliberal mandates. It has been cleaved from

94. hooks 2001.

95. Interview, July 7, 2015, New York, NY.

96. Beard 2014; Banet-Weiser and Miltner 2016. Of course, targeting women is not only an online phenomenon.

necessarily having anything to do with justice, as Sara Ahmed has so thoughtfully explicated. The political interventions discussed above, falling under loose headings of left or radical politics for their own sake, antimilitarism, and anticolonialism, have arguably less potential to be recuperated in this way. This makes them very important as critique, as imaginings of a fuller-throated invocation of justice in the context of technological development and practice. But at the same time, for that very reason they may be less likely to flourish in tech culture outside of these iterations, which are (and can likely only be) at the margins, not the mainstream, of tech culture.

One reason for this is that voluntaristic tech communities are still tethered to workplaces in significant ways. In the words of Johan Söderberg, "A programmer might freelance for a multinational company three days a week, spend two days as an entrepreneur of a FLOSS start-up venture, and in the meantime he [*sic*] will be a user of software applications, all of which are activities that feed into the capitalist production apparatus."[97] In other words, even off-the-clock use of software or innovation on one's own time are never very far from aligning with capitalist relations; at the same time, one is socialized to view one's own creative labors as authentically recuperable from instrumental workplace relations.[98] This constant blurring accounts for both San Francisco programmer Sam's impulse to bring up her own radical politics in a work campaign around web literacy as well as her immediate realization that a campaign at a corporate workplace was not an appropriate place to critique the prison-industrial complex in the United States. Contemporary creative economies are suffused with an "upbeat business-minded euphoria,"[99] which makes them resistant to the exercise of certain forms of critical agency. Many kinds of political language may be seen as alienating to colleagues, and to customers. Instead, Sam brought this up in a more hospitable setting, the unconference, but as a lament and a ritualistic proclamation of her politics.

It is worth revisiting Clara's statement, in which she observed a balancing act for people with left politics. As outlined above, she thought that in workplace settings, people with more radical left politics might latch onto diversity advocacy because the latter played well in corporate settings, musing that, "In my experience, one reason they sit together well is because the advocates often have social justice motivations but work in profit-driven organizations that aren't equipped, much less motivated, to do the right thing for its own sake. So advocates of course learn the language that will

97. Söderberg 2008: 6.
98. See Turner 2009.
99. McRobbie 2016: 27.

get an initiative funded, or whatever their goal is."[100] In other words, she felt that diversity advocacy in organizations might be a useful Trojan horse, allowing money or organizational time to be put toward political agendas that were harder to name outright in these contexts. On the other hand, as she readily admitted, these terms certainly circumscribed the scope of intervention. She humorously continued, "Ugh, I better quit [reflecting] before I start flipping tables!!"

Just as one is never outside of capitalist relations, "there is no vantage *outside* the actuality of the relationships between cultures, between unequal imperial and nonimperial powers," as Said writes.[101] In an interview, Christen, a German biotechnologist and activist for internet freedoms and women in STEM, pithily remarked, "I am just waiting for some hackathon organized by white people to make 'an app for Africa' or something."[102] Here she objects to hacker *solutionism* (and, slyly, to discourse that lumps African nations and peoples together as a monolith[103]). But she also points to one difficulty for hackers against colonialism: they are always already implicated in relationships of dominance and dependence. This is true for any person on earth, but it is uniquely true for technologists, who are so often called upon to execute political mandates under the guise of neutral technocratic intervention. It is difficult to ascertain exactly why no indigenous activists and few people of color showed up to the 2016 feminist hacking convergence in Montréal, for example, but the absence of these constituencies does underscore the potential difficulties of starting an anticolonial project centering on technology and specifically hacking.

Lastly, the Philadelphia hackerspace's polite tiptoeing around militarism illustrates the accretion of values and interests around electronics and computing that render them friendly to casual association with war, whether or not this is consciously articulated. Historian Paul Edwards has argued that the Cold War did not represent a new technological dawn but a continuation of World War II and the transference of a mythic, apocalyptic struggle onto a new enemy, placing computing at the center of this conflict.[104] Though this history is undoubtedly far from the minds of most hackerspace members

100. Personal correspondence with author, December 28, 2016. See Joseph 2018.

101. Said 1989: 216–17, emphasis in original. He continues, "When we consider the relations between the United States and the rest of the world, we are so to speak *of* the connections, not outside or beyond them."

102. Personal correspondence with author, July 2, 2015.

103. In an ironic twist, there is an app called "Africa isn't a country" for identifying "culprits" who perpetuate this discourse: see https://www.theguardian.com/world/2014/jan/24/africa-clinton.

104. Edwards 1996: 56.

celebrating a new drone design, the association has a sedimented history behind it that cannot be dislodged casually. This is why the DARPA debate at the HOPE Conference is so important. (Of course, there is nothing inherently liberating or peaceful about FLOSS: it does not matter if the drone that bombs your village was programmed using FLOSS.[105]) In his study of nuclear weapons scientists in North America, anthropologist Hugh Gusterson notes that there is a long-standing cultural commitment to working out issues of discomfort with militarism privately and alone, which he calls a process of "socialized individualism and collective privatization."[106] In other words, partly out of professional politeness, the ethics of weapons work are confronted alone, not in the professional realm. Again, the contiguity between norms of the workplace and of voluntaristic spaces has real consequences for what is politically articulable.

This is not to denigrate these ideations as goals for technologists—or for the society at large. Rather, the act of planting them here, in this medium, begs a question of whether they should be aimed at tech cultures or self-consciously cultivated in a wider milieu. The question of scale once again appears as a crucial issue; while it is a wholly understandable impulse to direct critique at one's immediate environment (especially an environment where one perceives acute problems), voluntaristic, local interventions may experience difficulty upending the edifices upon which technical cultures have been built. Inequality in tech cultures is an outgrowth of the fact that technology as a domain is vested with many social and political agendas that are not progressive, including technocratic, corporate, and capitalistic values that progressive diversity advocates directly contest. Technology cultures may—and do—breed resistant strains, but they may also prove to be especially hostile terrain for antiracist, anticolonialist, antimilitaristic, and just thought and practice to root and grow (as illustrated by the person who said she felt she was prevented from using a "moral framing" for her concerns). Technology is a uniquely challenging medium in which to plant and grow resistant politics. Yet there is also opportunity here: naming antimilitarism, anticolonialism, and antiracism as in bounds for care in technical communities—as feminist hackers have begun to do—can create a compelling starting point by shifting the terms by which tech culture is analyzed.

105. Thanks to Ron Eglash and Jonah Bossewitch for conversation on this point.

106. Gusterson 1996: 52.

7

Putting Lipstick on a GNU?

REPRESENTATION AND ITS DISCONTENTS

> Nowadays, it is claimed that the Chinese and even women are hacking things. Man, am I glad I got to experience "the scene" before it degenerated completely.
>
> —HACKER, 2008[1]

At the 2011 annual conference for the Python programming language (PyCon), held at a conference hotel in downtown Atlanta, GA, I immediately noticed how few women were there. Even though I had been intellectually prepared for a masculine-dominated environment, the experience was a little unnerving. Having hung around tech activism spaces for many years for research, I had become accustomed to being one of only a couple of women in a group of a dozen men. The software conference was a difference of several degrees of magnitude: it wasn't like joining a soldering meet-up at a squat and being a relative oddity, which might or might not be commented upon. This felt like walking into large, homosocial space like a barracks; I immediately felt that I did not belong, like everyone was looking at me, and I found myself stifling an urge to laugh (which was probably a reaction to how ludicrous I felt in this space and also masking discomfort). I kept a straight

1. "Unix Terrorist" hacker, 2008 (quoted in Coleman 2012b: 100).

face, pulled myself together, and, before long, adjusted enough to revel in my newfound ability to use the bathroom whenever I wanted without making an internal calculus over whether my need to relieve myself was greater than my distaste for standing in line or being late to the next session—an experience I had never had at a conference before or since.

This may seem an idiosyncratic or even frivolous way to mark the oddness of being a woman in the social space that is a men-dominated software conference. But it reveals several key points about this research site. First, it cannot be overstated that as a woman, one cannot enter such a space without standing out. This may be changing. It is not equally true in all sites. In fact, the very next year, when PyCon convened in Santa Clara, CA, the ratio of women to men was still not equal, but more women were present. This was a due to deliberate efforts by organizers.[2] It is logical to assume that people exhibiting other visible forms of difference from the dominant social identities in a given site might experience something similar. (For instance, as a white woman,[3] I cannot comment directly on the experience of people of color entering white-dominated spaces, but one imagines there are commonalities.) The extraordinary visibility one experiences may or may not create discomfort; this may or may not be overcome over time. Even not presenting in any hyperfeminine way (like many other attendees, I wore jeans and a rumpled button-down shirt), it was certainly hard to walk through a space crowded with men and not feel hypervisible. I did not experience leering, or any other inappropriate behavior, either at official conference sessions or in down time. But it was obvious that the eyes scanning over me—as they do at conferences, searching for friends, or associates one hopes to buttonhole—were taking longer to process my presence or sort me into a category ("not known, not important," I would guess).

Visible, marked difference may be a source of comment, too. It was difficult to not take ordinary small talk inquiries (all of which were entirely polite and mundane) about who I was or what I was doing at PyCon as specially interrogating my presence, at least somewhat. In a way, being there as a true interloper, a social science professor interested in the dynamics of the community, both broke the ice and gave me cover. I would imagine that being a first-time attendee, a woman, and a programmer could make similar

2. In 2010, Google had given money to PyCon to bring in women attendees and speakers. In 2011 and 2012, people commented on how that year had been a higher-water mark than 2011 was, where the number perceptibly dipped back down.

3. I am speaking as a cisgendered woman. For people with nonconforming gender presentations, a similar experience might happen entering many mainstream cultural spaces.

inquiries feel like one's bona fides or right to be there were under examination, depending on the tone and circumstance. (Women programmers routinely report being assumed to be at software conferences as someone's girlfriend, or because they are in marketing, not because they have technical roles.[4]) Being such an obvious outsider, I was received quizzically, but my presence was not challenged. That said, it would have been relatively easy for dynamics to tip in a direction that felt less welcoming, and if that had happened, it would have been hard not to wonder if who I was (or was perceived to be) had something to do with that sort of interaction. It is also possible that affirming that I was *not* a programmer confirmed subtle biases about women in computing, and made me easier to place—that, ironically, being a total outsider was easier to explain.

All of this is preamble to the main topic of this chapter, which is how—in a climate of burgeoning awareness of diversity issues and advocacy to change the ratios of open-technology participation—community members grappled with issues of social position and identity. I was at the 2011 PyCon precisely because I had identified the Python community as one where diversity advocacy was recently picking up steam; a network of PyLadies or PyStar groups to support women, trans, and nonbinary folks learning Python had begun to form in a variety of cities in 2010–2011. Though most of the technical talks at PyCon (which were, of course, the vast majority) were beyond my ken except at the most conceptual level, there were multiple formal presentations about diversity on the conference program; moreover some more informal "birds of a feather" sessions formed too. I quickly identified a handful of people who seemed like ringleaders for these topics, and attached myself to them. Probably not least because of their reflective awareness of being welcoming as a tenet of diversity work, they were kind, and willing to be ambassadors of PyCon for me.

Background on Social Identities and Technical Cultures

This chapter focuses on how advocates for diversity in open-technology cultures define the identity categories around which they rally. It is evident that they hope to deepen an association between certain categories of social identity and certain forms of technological belonging (as well as challenge

4. Ciara Byrne, "The Loneliness of the Female Coder, September 11, 2013, https://www.fastcompany.com/3008216/tracking/minding-gap-how-your-company-can-woo-female-coders. Though any given account is anecdotal, there are a lot of them.

other associations). Contemporary social studies of technology treat technology as neither wholly socially determined nor as conforming to or flowing from an internal rational logic. Technologies and technical practices are understood as durable but not immutable assemblages of social relations and technical artifacts. Technology—more specifically, people's relationships with machines—is an important site of identity construction.[5] I conceptualize the people in these voluntaristic open-technology communities not as people who simply work and play with computers and electronics (and crafts), but as people who actively construct identities around their work with these artifacts. They hold a closer relationship to technology than average users.[6]

The people discussed in this book have encountered and contested longstanding patterns of how technological affinity intersects with social identity and social position. Diversity advocates in open technology are openly struggling with the legacy of computing in North America and Europe as a pursuit that has historically been linked with white men, often from privileged educational and economic backgrounds.[7] (Chapter 5 addresses employment relations and social class more directly than the current chapter.) This is partly due to a legacy of technical enthusiasm that formed in the early twentieth century around radio tinkering, which enforced both electronics hobbies and white-collar technical professions as the province of white men.[8] Computer programming and hardware tinkering is a later development, but computer enthusiasts represent continuity with earlier forms of technical masculinity, as opposed to a radical new class of people birthed by the introduction of computers. That said, devotion to computing has some distinct traits: the geek stereotype was codified as a concept and a modern vocabulary word around computer enthusiasts.[9] Scholars have noted that "in spite of the possibility of emancipation from corporeal realities imagined by early theorists and boosters of new media and cyberspace, bodies and social positions are anything but left behind in relationships with computers . . . it is still the case within the so-called high tech and new media industries that 'what kind of work you perform depends on how you are

5. See Orr 1990; Douglas 1987; Turkle 1995; Haring 2006.

6. Dunbar-Hester 2014; Haring 2006.

7. Kendall 2002; also see Ensmenger 2010a and Dafermos and Söderberg 2009 on class position.

8. Douglas 1987; Haring 2006; Oldenziel 1999.

9. Dunbar-Hester 2016a.

configured biologically and positioned socially.' "[10] In other words, social context and position, including gender, matter greatly as we consider who participates in technical practices, and who possesses agency with regard to technology, both historically and in the present.

Gender is a relational system for sorting people, which intersects with other systems for social sorting.[11] Gender as a category has existed across history, though the particular content of what constitutes, for example, masculinity, femininity, or binary-straining gender formations, has varied over time and across cultures. Gender stands as a useful category for analysis because it is a constitutive element of social relationships based on perceived differences between people of different genders, and it is a primary way of signifying relationships of power.[12] Gender identity is not something that is given, it is something that is constantly constructed and remade: "the 'doer' is variably constructed in and through the deed," in the words of Judith Butler.[13] Identity is not endlessly fluid merely because it is performative or iterative.[14] Rather, identity is constituted through performance, through materiality, through practice, through social relations, through signification and representation. Scholars have shown that "technology [is] both a source and a consequence of gender relations."[15] Gender structure and identity[16] are materialized in technological artifacts and practices, and technology is implicated in the production and maintenance of a relational system of gender.[17]

As Ron Eglash has explored, the whiteness (and masculinity) embedded within geek identity means that members of other racial categories have often had to reject or recast hegemonic geekiness in order to find a comfortable place for themselves within technical subcultures. He specifically discusses Asian American "hipsterism" and Afrofuturism as racially

10. Sara Diamond quoted in Suchman 2008: 149.

11. Scott 1986.

12. Scott 1986: 1067.

13. Butler 1990: 142.

14. Butler 1993.

15. Wajcman 2007.

16. It is widely acknowledged that gender occurs not in isolation but within a matrix of factors that affect social identity, which include class, nationality, ethnicity, and race.

17. In spite of the attention given to identity, I in no way mean to discount social structure (along with gender symbolism) as sites of gender production (Lerman et al. 2003: 4). See also Faulkner 2007. It is tricky business, but both individual agency (individuals "doing") and social structure contribute to how gender is constituted. Social structures and institutions including patriarchy, capitalism, and colonialism may impose consciousness of group membership on individuals (Haraway 1991a: 155) race and gender are conditions as much as identities.

recoded appropriations of technology. In other words, these are nondominant groups' strategies for extending technological affinity to nonwhites in ways that challenge (but do not totally evade) the hegemonic whiteness of geekdom. (I invoke race and racial categories here under the terms proposed by proponents of critical race theory, who hold that "race and races are products of social thought and relations. Not objective, inherent, or fixed, they correspond to no biological nor genetic reality; rather, races are categories that society invents, manipulates, or retires when convenient."[18]) Jessie Daniels calls for greater attention to race and racism in internet studies, arguing that too little attention has been paid to the power dynamics and social structures that have moved racism online (and often transformed it there in new and troubling ways, including globalizing it).[19] Quoting Carolyn de la Peña, she argues that racism has "whitened our technological stories," which are typically unable to reveal the whiteness of not only internet cultures but even the scholarly interrogations of them.[20] The present chapter attempts to account for whiteness even in unmarked forms, and to explore how racialized notions of technological participation have led to both named struggle and unidentified blind spots for those who would diversify open-technology cultures.

By definition, identity categories are constructed. One benefit of using identity as a means to understand relationships with technology is how it may help us to get at parts of human experience that are moving targets, slippery, constructed, and yet real.[21] Structural factors certainly contribute to individuals' and groups' relationships with technology. But here I interrogate identity categories as they are introduced and interpreted by diversity advocates, not because structural factors are unimportant, but because forms of social belonging experienced around relationships with technology are salient for diversity advocates. It goes without saying that there are potential limitations in understanding how race and gender categories become interwoven with technical identities as primarily about identity work; technological cultures are made of (and with) much more than just identity work. (Arguably, though, identity categories feel somewhat more within actors' control than structural factors.)

18. Delgado and Stefancic 2001: 7. Racial categories (and even the concept of race itself) are situated in time and place.

19. Daniels 2013: 16; see also Nakamura 2007; Nakamura and Chow-White 2011.

20. Daniels 2013: 17.

21. Dunbar-Hester 2014: 24.

Taking the intersection of social identity and experiences with tech as a starting point, this chapter describes how diversity advocates have attempted to grapple with the "who" of diversity advocacy. In other words, which kinds of subjectivities and social identities do advocates believe are missing from tech? How do open-technology communities conceive of the diversity they hope to usher in? What is at stake in these contestations over representation? What is at stake in thinking of this as a matter of *representation* in the first place? This chapter argues that representation as a goal has limits as a project of empowerment, as noted by scholars of postfeminism and race.[22] Representation as a producer or a consumer of a given technology is not necessarily indicative of a change in social power or social status.[23] If diversity advocates in open-technology communities are serious about changing how social power is configured, it might serve them well to focus on goals beyond representation in technological participation. While it is easy to conflate technological representation and social power, they are not interchangeable and should not be mistaken for one another.[24]

Gender, Trouble

Within diversity advocacy writ large (including not only open-technology communities, but industry and government discourses), gender diversity is commonly held as a primary goal. In fact, *diversity* is frequently used as a metonym for *gender diversity*, which, in turn, is reducible to *more women*. In other words, *diversity* is often almost a shorthand for *more women*, and *women* can stand in for *diversity*, too; each collapses to mean the other. For example, some advocates for diversity claimed that collapsing *diversity* into *women* was justified because improving women's participation was a surefire way to increase other, unmarked forms of diversity overall. For example, at a talk at the 2011 PyCon conference, a speaker said, "Gender diversity can broaden out to other kinds of diversity."[25] Diversity advocates in open-technology communities wrestled with this framing of *diversity*, sometimes

22. Banet-Weiser 2012; Gray 2013.

23. Eglash et al. 2004: xv.

24. Dunbar-Hester 2014: 188.

25. Fieldnotes, March 11, 2011, Atlanta, GA. Another typical example of this sentiment, found in the comments on a blog post about diversity and culture in open source, may be seen here: "I bet that many of the changes we would make to improve participation by women would also improve participation by other under-represented groups!" "I've Been a Terrible Person (and So Have Most of You)," May 21, 2012, https://mjg59.dreamwidth.org/11799.html?thread=367383.

aligning their efforts with it, and at other times challenging it along multiple lines.

As noted multiple times in this book, efforts to promote women's inclusion in open technology have ramped up in recent years. We have seen the founding of groups like LinuxChix (founded ca. 1998), Debian Women (the Debian operating system project, ca. 2004), Ubuntu Women (2006), the Geek Feminism project (ca. 2008), and, more recently, PyLadies (from the Python computer language community, 2011) proliferate, and the list goes on. One mundane example of this highly self-aware turn toward inclusiveness[26] can be seen in the following email exchange. One person (with a masculine username) addressed an electronic mailing list, Womeninfreesoftware, hoping to recruit women to FLOSS projects in which he was involved:

> I had a look at the projects I'm directly professionally involved in—[project A] and [project B]. And, well, they're pretty much your typical FLOSS sausage fests [men-dominated spaces], I'm afraid. We do actually have a few women involved, but they're all [company] employees [who are paid to be there]; *on the volunteer side, it's all men so far.*
>
> So I'm hoping to encourage people—women in particular—reading this list to come and get involved with [project A] and [project B].[27]

This quote exemplifies a project participant reaching out to women and in so doing, demonstrating that these projects were aware of diversity issues and making an effort to be welcoming. Another list subscriber replied, "Thanks, [Name], for taking the time to make that bid for participants in your project. *It was exactly what the world actually needs*[,] much more so than almost any other single action."[28] These quotes illustrate the typical, mundane framing of gender diversity as inclusion of women in free software projects (all post-2006 initiatives should be read in part as a reaction to the 2006 EU policy report by Nafus et al. showing a minuscule rate of participation by women). Reaching out to women was seen as a straightforward, and appreciated, way to foster the inclusion that diversity advocates prized.

A major concern of diversity advocates, which runs throughout this book, was how to rethink the gender codes surrounding hacking, coding,

26. Paulitz et al. (2012) have argued for heightened reflexivity on the part of the researcher when conducting research within a field already shaped by gender equality politics.

27. Email, [A—] to [Womeninfreesoftware], September 28, 2009. Emphasis added to show that women voluntarily showing up to men-dominated technical spaces is sometimes a tough sell.

28. Email, [K—] to [Womeninfreesoftware], September 28, 2009, emphasis added.

and computing. In particular, advocates wished to challenge tacit or explicit associations of masculinity with coding and hacking, which they believed could begin to undo the "sausage fest" climate of open-technology projects. This was a constant refrain. It is so ubiquitous that outside of this chapter I rarely comment in depth on the embedded assumptions about gender that diversity advocates were sometimes exposing in order to challenge, and sometimes recapitulating. But here they are of central concern. It is worth considering in detail some of the wide-ranging claims that constitute how diversity advocates were framing and responding to these issues.

Why did open-technology practitioners identify advocacy around gender and participation as a primary site of intervention? How did they understand the lifeworld of open technology, and in what ways did they feel it needed to be changed? One example of the lifeworld of open technology that advocates hoped to dispute can be seen in the following quote. During the dispute in the Philadelphia hackerspace about whether to host an event featuring sex toy hacking (chapter 3), one member wrote the following statement as he expressed his reluctance to host the event:

> A lot of the hackers here at the space are the *Make* Magazine/Instructables type, not the Julian Assange HOPE-conference attending type, or even the kind that cares much about a global movement of hackerspaces. *I'm not sure what category dildo hacking falls in.* . . . For a lot of people, DIY has to do with a sort of Father-son nostalgia for *Popular Science* magazine, model airplanes, and so on, so even if we don't actually get too many kids in the door there is a lot to be said for promoting that image.[29]

This comment is quite illuminating. The hackerspace member allows for a range of hacking practices, first invoking an independent producer ideal, exemplified in *Make* magazine (though Maker Faire and much of DIY actually reproduces consumer culture[30]). He next refers to a more politicized, Wikileaks-esque, information-wants-to-be-free sort of hacker.[31] Then, he correctly signals the heritage of DIY as a way to impart technical affinity to boys. Notably, while he allows for different motivations for hacking, they are all tied to masculinity, which is marked, and whiteness, which is not. In other words, part of what diversity advocates struggled to address was that even as *hacking* itself might be malleable enough to encompass a range of

29. Email,—to [HackMake list], February 4, 2011. Emphasis added.
30. See Gelber 1997; Sivek 2011.
31. See Golumbia 2013a; Coleman 2015; Winner 1986.

politics and practices, the default hacker was under all of these scenarios a man or a boy.[32]

This association between hacking technologies and masculinity was thus a prime site for advocates to intervene. Over and over, advocates discussed what they perceived as a tacit, entrenched embrace of masculine identity performance in the practices of collaboration around software and hardware. Many early studies of computer geeks have argued that one of the affective pleasures people find in computing is mastery. Writing in 1996, Sherry Turkle observed that "[Hackers] are passionately involved in mastery of the machine itself."[33] (This passion differentiated highly skilled and enthusiastic user groups like hackers from ordinary users who were interested in running applications, but not necessarily trying to get inside the machines' code or hardware themselves.) This is consonant with the construction of masculinity around control over technologies that preceded computing.[34] In computing, the exaltation of mastery occurred in part through competitive dynamics, often evinced in fleeting moments of interaction, but in aggregate constituting a cultural norm (to which many computer enthusiasts were exposed quite early in life[35]). Socialization practices reflect these norms: old-timers in particular often recalled being hazed while learning to program. Even when one is not being hazed per se, there are decades-old geek norms to "read the fucking manual" (RTFM) before asking questions, which discourages people from displaying uncertainty. Joseph Reagle writes that the RTFM norm can provide a positive incentive toward self-cultivation: one will be valued if one strives "to learn [to code], to write a useful utility, and [then] to share [one's] learning and its fruits with others."[36] Less positively, of course, this directive to independently cultivate and exhibit mastery can also be used to intimidate, to shame, and to turn off newcomers who are for whatever reason unwilling or unable to flourish under these conditions. This effect has been celebrated in mainstream open source. Eric Raymond,

32. "Dildo-hacking," of course, references sexuality; though the poster seems to be objecting to the presence of adult sexuality in a space that is also for "father-son nostalgia" and possibly actual "kids," it is worth raising the question of whether *all* sexuality is troubling, or whether discussions of sexuality that center on the perspectives and experiences of people who are not straight men are particularly distressing for the poster.

33. Turkle 1995: 32. Here I sidestep Turkle's arguments about "hard" and "soft" mastery and how they may or may not correspond with gendered programming styles.

34. Wajcman 1991, chapter 6.

35. See Margolis and Fisher 2003: 35–43 for a discussion of gender and competitive dynamics among middle- and high-school students.

36. Reagle 2016: 692. See also Coleman 2012a: 107.

open-source evangelist and author of the famed essay "The Cathedral and the Bazaar" writes,

> We're (largely) volunteers. We take time out of busy lives to answer questions, and at times we're overwhelmed with them. So we filter ruthlessly. In particular, we throw away questions from people who appear to be losers in order to spend our question-answering time more efficiently, on winners. If you find this attitude obnoxious, condescending, or arrogant, check your assumptions. We're not asking you to genuflect to us—in fact, most of us would love nothing more than to deal with you as an equal. But it's simply not efficient for us to try to help people who are not willing to help themselves.[37]

This ethos has also been subject to pushback; Reagle notes that Wikipedia reminds its community members to "not bite the newcomers."[38] Nonetheless, these dynamics are well documented as long-standing tenets of computing culture and especially open source, and it is not necessary to rehearse them here beyond a basic adumbration.

These dynamics conform to a normative masculinity, where competitive displays and mastery are understood as virtues. Scholars have argued that geek masculinity is in some ways distinct from hegemonic masculinity,[39] but it still trades in "othering" women.[40] David Bell writes, "Geek masculinity [is] a complex negotiation of outsiderhood and privilege, even outsiderhood *as* privilege, which sometimes suggests possibilities for gender inclusivity, but at other times reinstates rigid gender boundaries and hierarchies."[41] It is worth noting that *masculinity* is a set of (contingent and historically malleable) characteristics that have been laden with the power to signify men and maleness, but do not necessarily inhere to men. Nor is it the case that women (or nonbinary folks) may not possess or exhibit traits that are gendered masculine. Rather, signifiers of masculinity and femininity are components of a symbolic repertoire that support a cultural system of gender.

37. Quoted in Reagle 2016: 697.

38. Reagle 2016: 697. Reagle notes that this has not been all that successful, as Wikipedia still has a reputation for being "fighty."

39. Scholars have suggested that nerd identity is, in fact, *unsuccessful* masculinity, "threatening the masculinity of its participants" (Eglash 2002: 51). The distinction between *geek* and *nerd* is illustrative: a geek may or may not elect this identity or revel proudly in it, but a nerd *always* is a nerd involuntarily (Eglash 2002: 61n1). Both the masculinity and failure embedded in geek/nerd identity could be responsible for women and girls' ambivalence toward it.

40. Kendall 2002: 87. See also Wajcman 1991, chapter 6.

41. Bell 2013: 79.

Contesting Gender Essentialism

With this as introduction, how have diversity advocates dealt with normative masculinity as a component of open-technology cultures? The answer to this question is not straightforward, and it illustrates many pitfalls for advocates. One major stumbling point, which seemed all but impossible to resolve, was how to challenge normative masculinity as a cultural default without invoking normative femininity as its opposite. This played out in a variety of ways.

On the list where the person had lamented the "sausage fests" in his current open-source projects, list members discussed making a logo to represent the list itself. One list subscriber proposed, "If we took the picture of a GNU used by FSF [Free Software Foundation] . . . added lipstick, eye shadow, and mascara, replace the beard by a string of pearls, and replaced the horns by a feminine hat, with a flower sticking up from the hat, I think that would convey the idea."[42] The "GNU symbol" is the logo of a Unix-like, Linux-related operating system, a line drawing of the antelope-like gnu, replete with beard and horns as described in the email (figure 7.1). In other words, the subscriber proposed adorning the GNU gnu with normative markers of femininity.

Responses to this suggestion registered immediate discomfort. One person commented, "I . . . am not a big fan of this idea. Most women in free software do not adhere to traditionally feminine styles of dress/grooming—I have seen very few wearing makeup let alone pearls at free software events—and I think this sort of appearance would be alienating to many of us."[43] The original poster agreed with this ("You're right. . . . Most of us don't dress over-the-top feminine. I certainly don't."[44]) and added that the original suggestion was intended to be a humorous way of depicting women in FLOSS. Posters to the list struggled with how to signal the presence of women without falling back on representations of normative femininity that many of them found "alienating." They also touched on race, as one commenter wrote, "I think the gnu is more appealing than the wasp-y [white Anglo-Saxon protestant] noses and dainty lips [in some other ideas for a logo]."[45]

Along similar lines, in a question and answer session following a presentation called "Diversity in Tech: Improving Our Toolset" at the 2011 PyCon

42. Email, [M—] to [Womeninfreesoftware], September 24, 2009.
43. Email, [K—] to [Womeninfreesoftware], September 24, 2009.
44. Email, [M—] to [Womeninfreesoftware], September 24, 2009.
45. Email, [A—] to [Womeninfreesoftware], September 24, 2009.

FIGURE 7.1. GNU logo, GNU Project, ca. 2009. Used with permission under the Creative Commons Attribution-ShareAlike 2.0 License.

meeting, normative assumptions about femininity caused significant controversy. The presenter, a thirty-something white woman who worked as a software engineer in a large Silicon Valley firm, offered an overview of research that addressed "diversity in the workforce." Beginning with a description of "gender diversity" (meaning women) in the tech industry, the presenter said there were things that made women more or less likely to participate in tech fields. She described a hypothetical example where a woman approached a recruiting manager, who was a man, about an engineering job, and was met with a brief moment of disbelief that she was the applicant (because of her gender). The presenter claimed that research showed that expectations have an effect on performance, suggesting that the woman engineer might be undermined by the manager's expectations: "The engineering manager's expectations would have impacted how he judged her, and could conceivably have impacted how she performed in a job if she had taken that job."[46] She pointed out that there was "no malice on the part of the engineering

46. Fieldnotes, March 11, 2011, Atlanta, GA.

manager," but "if you're not watching what you're doing, you can actually materially damage the ability of other people to succeed alongside you."[47] She then told the audience that "working in a diverse workforce is hard. It's harder than working with people like you. . . . You will be more likely to face people challenging you." In the end, though, she claimed that any potential strife had a payoff: "Conflict and challenge [are] how we arrive at interesting, creative solutions to problems."[48] Her concluding point was that, "Diversity is hard, but if you work hard at it, you can end up ultimately with a much better situation."[49]

On the one hand, this presentation was utterly banal. It signaled the importance of diversity, invoked social science research, and argued that embracing diversity was, while somewhat challenging, likely to yield benefit to firms and workers (and implicitly to consumers). This line of argument is eminently familiar from many corporate, higher education, and government discourses about diversity. It was certainly presented as something no reasonable person could disagree with. And yet somehow, it was apparent that the presentation had caused the temperature of the room to rise. There were a few possibilities as to why. First, the presentation had, knowingly or not, the tone of a human resources–driven (and mandatory) workplace training session, on which some attendees later commented disparagingly.[50] While it did not address topics like harassment, the tone was fairly dour. It occurred to me that given that the audience members were all there voluntarily (this was not a plenary session, so they could have attended several other concurrent talks instead), many of them possibly were not there for a stern lecture making the point that diversity is a virtue. Diversity advocacy can be a volatile topic in the "sausage fest" environment of computing and has been interpreted as a threat to the "clubhouse" atmosphere of programming workplaces and (especially) FLOSS projects. This presentation probably did little to enhance the notion that that diversity work was anything other than a slog, or that diversity conferred benefits that intersected with the affective dimensions of computing prized by many. (This stood in contrast to another presentation earlier that day, given by advocates for diversity in open source whose focus was more on voluntaristic projects. Entitled "Getting More Contributors (and Diversity) through Outreach," advocates

47. Fieldnotes, March 11, 2011, Atlanta, GA.
48. Fieldnotes, March 11, 2011, Atlanta, GA.
49. Fieldnotes, March 11, 2011, Atlanta, GA.
50. Fieldnotes, March 11, 2011, Atlanta, GA.

there emphasized finding out why people were involved and whether they were having fun.[51])

The "Improving our Toolset" presentation became volatile during the Q&A following it, along lines that are perhaps surprising. Audience members had lined up to ask the speaker questions that presented scenarios they had encountered and asked what could have been done differently. One fellow, a white man in his thirties or forties, began his question by remarking that he had been married recently. He said that his wife had some talents that included "communicating, worrying, and correcting my grammar and spelling mistakes."[52] He said the presentation had caused him to think about how if men and women were "fifty-fifty" in programming, coding might be better documented and contain more "case statements." (These are conditional control structures in programming that allow for a selection to be made depending on whether a clause directs one choice or another; in rough terms, analogous to "if this, then that," or especially, "if this, and this, then that." In other words, he was referring to thinking about outcomes resulting from multiple layers of contingency, perhaps born of a cautious mindset; in his words, "is this gonna happen, is this gonna happen, is that gonna happen.") He mused that teams might be stronger if they were composed of people with complementary skills, which he saw as correlated with gender. At this point, an audience member, a white man in his thirties from Minneapolis who was also in line to ask a question, had an outburst, shouting, "COME ON, MAN!! Are you serious?!??" and stormed out of the session. The speaker gingerly replied, asking if the querent was suggesting that "code would be better documented, and would it cover cases that might go wrong" if his wife were a programmer. She said she had not seen research that indicated that this was true, and then added diplomatically, "I find that it's good to be really careful about making gender assumptions."[53] The Q&A then turned toward some of the issues that this question had raised. The next person to speak, a white woman with a northern European accent, challenged the previous querent. She pounced on the suggestion that inherent gender differences guided people in different domains of action, and said, "stereotypes don't help, even if they're complimentary," at which point many people spontaneously applauded. She also said, "What about women in politics? People think it would be so much better if the world were

51. Fieldnotes, March 11, 2011, Atlanta, GA.

52. Fieldnotes, March 11, 2011, Atlanta, GA.

53. Fieldnotes, March 11, 2011, Atlanta, GA.

run by women, there would be no more wars—but ever hear of Margaret Thatcher? . . . When you generalize, even if it's well-meaning . . . [I would say] thanks for the compliment, but no thanks."[54] As the original querent reacted to the presentation and attempted to map its topics to his own observations, he invoked gender stereotypes of women as caring, cautious, and communicative (which were, he averred, different from his own approach, an implicitly masculine one). He was no doubt surprised by the reaction to his question, which, it seemed, was sincere, and was not in any way overtly denigrating toward women. But immediately, both the presenter and two other audience members resisted this essentialism—more and less forcefully.

Though the fellow asking the question that invoked his wife was undoubtedly taken aback by these reactions, diversity advocates were primed to be impatient. Certainly, *wife*, *mother*, and other roles gendered feminine are often used as shorthand for *nontechnical people*;[55] language habits have been a way of entrenching the notion that women are outsiders in computing. Second, though, this fellow's question and the ensuing discussion illustrated that to advocate for women in open technology is to have ideas about who women are and why their presence is a good thing. Provisionally agreeing on diversity (or more women) as a virtue is, of course, not the same as forming consensus about these matters. These are thorny issues, rife with identity politics, issues of representation, and, as seen here, questions of essentialism. In the case of people even casually steeped in gender studies, which includes no small number of people in open-technology communities, to promote women in technology was to fairly immediately run afoul of gender stereotypes. The fellow who asked the question invoking his wife was either not familiar with or in disagreement with common critiques of gender essentialism. His misunderstanding of the issues elicited impatience in part because while advocates tended to agree that essentialism was a problem, how to combat it was far from clear, both among advocates themselves and when they were facing wider open-technology communities. And advocates did not necessarily produce any converts to their way of thinking in this interaction.

Later, in an impromptu discussion about the presentation, people criticized the presenter for turning off audience members. They said she had struck the wrong tone for the developers and community managers who

54. Fieldnotes, March 11, 2011, Atlanta, GA.

55. A mundane example of this sort of speech is the title of this (Mother's Day!) blog post, "The Internet of Things, So Mom Can Understand" (Karel 2015).

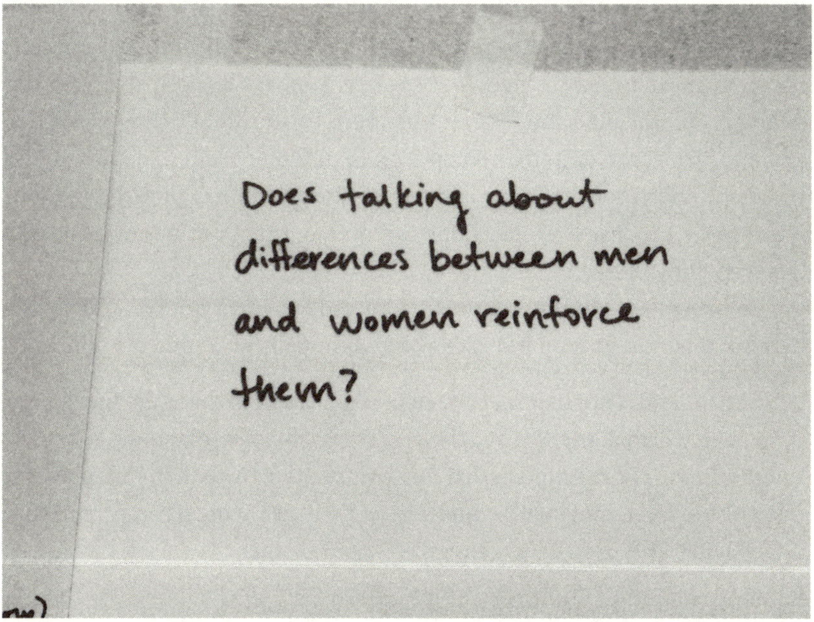

FIGURE 7.2. Session proposal, "Does talking about differences . . . ," Washington, DC, July 2012. Photograph by the author.

were there, most of whom were not hiring managers, and who were probably already "on the train," accepting of diversity as a good, at least in the abstract, and not in need of that kind of "remedial information."[56] They did not, however, address how the explosion over gender stereotyping might have puzzled or alienated people who were less conversant in critiques of gender essentialism.

But more sympathetically, taking the measure of the GNU and PyCon examples, it becomes evident that arguing for gender diversity without promoting gender essentialism was a dilemma for diversity advocates. Advocates themselves acknowledged this in many ways. At an unconference devoted to "women* in open technology," one person proposed a conference session to talk about this double bind: how to promote women in open technology without reducing people to stereotypes. She expressed this on a note, taped to a wall where people posted various ideas for sessions (figure 7.2):

Her session topic centered on a question, "Does talking about differences between men and women reinforce them?" The simple phrasing of the

56. Fieldnotes, March 11, 2011, Atlanta, GA.

question belies a tangled thicket of issues that social scientists and theorists of gender in culture have not resolved; there is certainly even less consensus about these topics in the wider society. She was wary of stereotyping members of masculine and feminine genders, especially allowing a focus on gender *difference* to amplify perceptions of difference and thus potentially to actually construct or deepen difference. The question is almost poignant, in that one can sense both her hope for moving past divisions and her fear of entrenching them.

Later that day, Miriam, a woman who had for years worked and volunteered in technology communities, offered thoughtful comments. She said:

> The problem is not that men are in workplaces [and therefore that] more women will make workplaces better. The problem is masculine behavior. Women [too] can emulate that, act competitive. And then that makes it even worse, [people will] comment on them as [being] "catty" and [that behavior will] be used to generalize about women.[57]

This comment raises many issues worthy of scrutiny. It presents the belief that gender differences are not inherent; if women can act "masculine" and competitive, it stands to reason that men can act in less stereotypically masculine ways as well. She rightly notes that technical cultures are often competitive and links this to masculinity, but does not think that "just adding women" will necessarily change the cultural norms.[58] Further, she worries that placing women in environments where masculine norms hold sway will set them up for failure; even if they succeed in aping in those patterns of behavior, they may then be penalized for not being appropriately feminine: competitiveness, a virtue in men, in a woman is perceived as cattiness, a vice. At a subsequent meeting of the same unconference, conversation was devoted to what one person called the "likeability paradox": she said that for women being liked and being perceived as competent were inversely correlated, especially in men-dominated fields.[59] She said that this meant that "women have to choose between being liked and being powerful." Social science research suggests that there is truth to this claim, and that at the

57. Fieldnotes, July 10, 2012, Washington, DC.

58. In earlier ethnographic research on radio activism, I have observed this assertion being borne out: young women entering a masculine-dominated volunteer technical community adopted masculine, competitive norms of the group, as witnessed by another volunteer (Dunbar-Hester 2014: 64–65).

59. Fieldnotes, June 6, 2013, San Francisco, CA.

very least this terrain is much more challenging for women.[60] And in an interview, another woman said that in an earlier period in her career, "I had worked hard to emulate assholes."[61] In other words, in domains where competence was strongly coupled with technical masculinity (performed through competitive displays centering on technical knowledge), placing technically competent women in this environment would not necessarily make the environment "less masculine," nor see the women rewarded for their competence.

All of this is to say that there was a high degree of reflectiveness about gender and a strong current of questioning essentialist tropes among these advocates for diversity. Even when engaging in conversations about differences between men and women, people usually tried to stake out territory that allowed for differences among members of a given gender. One person expressed this when she said at an unconference meeting, "When people say 'women [have this in common; are this way],' it should not mean '*all* women are [this way; have this in common]."[62] They also routinely allowed for the possibility of common characteristics in people of different genders. The above advocate's position that "adding women" was a less satisfactory solution to gender trouble in technical communities than titrating out some of the communities' masculinity was echoed in different ways. Though subtle, this is a notable distinction. Especially in environments that were critically oriented toward gender, which many of these were, people did not view femininity as a polar opposite to masculinity, and they also did not believe that all women were necessarily especially feminine.

Not uncommonly, people suggested imbuing environments with subtle cues that subjected the alpha geek competitive masculinity common in open-technology communities to subversion or redirection. One person said that in her workplace, new staff, regardless of gender, were given colorful arm-warmers and a brightly-colored toy.[63] Another person suggested that it might be beneficial for members (of all genders) of her hackerspace to all go out and "get their nails painted with the hackerspace's colors."[64] One person in an interview said that she had seen people use glitter at events to lighten them up; she claimed that in places where an all-black hacker aesthetic was dominant, festooning people with glitter was a way to inject a

60. See for example Heilman and Okimoto 2007.

61. Interview, Liane, July 24, 2014, San Francisco, CA.

62. Fieldnotes, June 6, 2013, San Francisco, CA.

63. Fieldnotes, June 6, 2013, San Francisco, CA.

64. Fieldnotes, June 6, 2013, San Francisco, CA.

feeling of "color and celebration and comfortable vibes."[65] Notably, no one here was suggesting that glitter, colored armbands, or hackerspace-themed nail polish would *feminize* masculine spaces, nor was this their goal. In fact, these interventions can be read as resistance to the likes of "pink technology," a marketing strategy to promote girl-oriented consumer electronics, as described by Mary Celeste Kearny.[66] Rather, they were intended to queer or attenuate normative masculinity: the toy and the arm-warmers in bright primary colors reference childhood as much as anything else, and glitter and group nail polish evoke carnival or drag. (Another person described a community of coders who focused on both technical development and maintaining an inclusive environment: "[they] sport a lot of rainbows and glitter in a kind of supercampy anti-machismo."[67]) It is perhaps debatable whether other people outside of these subcultures—or even everyone within them—would read these signifiers the same way. Nonetheless, it is apparent that for these actors, these strategies represent attempts to decenter masculinity without replacing it with normative femininity, of which they were critical and with which many of them were uncomfortable (as mentioned in the "lipstick on the GNU" discussion above).

Genderfuck: Queer, Nonconforming, Otherwise

This becomes especially apparent when we consider other aspects of gender that were swirling around in this milieu. In the discussion about the GNU, list members also identified another issue. One person commented, "I think the question of gender identity goes deeper than 'do we all wear pearls here at [Womeninfreesoftware]' to, are we really limiting our reach to 'women' or is there also room for gender queer techies who don't identify with the gender binary?"[68] In other words, using normatively feminine markers of identity to represent women in FLOSS was problematic for two reasons. First, these markers invoked and threatened to reinscribe a version of femininity that many geek women did not relate to, as discussed above. Second, they undermined a commitment to gender diversity salient in open-technology circles, where the prevalence of nonbinary- and trans-identified people seems relatively high (or is, at least, visible and vocal). In other words, for

65. Interview, Perez, November 4, 2016.

66. Kearney 2010. Thanks to Bryce Renninger and Laura Portwood-Stacer for discussion on this point.

67. Personal correspondence,—to author, December 28, 2016.

68. Email, [P—] to [Womeninfreesoftware], September 24, 2009.

many, *gender diversity* did not stop with "more women." Many projects and hackerspaces with a commitment to gender inclusion explicitly address and include people who identify as genderqueer, nonbinary, and so on. One representative example is from a feminist hackerspace, which describes its community as "intersectional feminists, women-centered, and queer and trans-inclusive."[69]

There is a likely prehistory here, which I can only suggest. Preceding the fairly recent bursting of trans presence into the cultural mainstream, trans, genderqueer, and nonbinary gender identity in many tech circles was already somewhat established. There are a variety of factors at play, certainly, but one stands out as worthy of sustained consideration. Whether or not people are self-aware about this connection, there seems to be some correlation between people who are drawn to hacking and tinkering and people who are drawn to questioning gender as a system. Many people make this connection self-consciously, as did Salix, a queer-identified, twenty-something person who had cofounded a makerspace in the Pacific Northwest, who said,

> Why are there so many queer geeks? To me the word "hacker" is about anyone who wants to tear down a system. It's the same with [my] gender and sexuality—tearing down a system and seeing where I fit in it. I wonder if it's something in brain chemistry [where people are "wired" to be attracted to tech and to question gender]. [Hackers] also have the experience of construction of self online. You play with construction of self, you play with sexuality. . . .
>
> Through the technology we're building now, we are expressing ourselves . . . [there are opportunities] to finally get into the nuances of your personality . . . we are seeing [new ways for] how people express themselves.[70]

Playing with selfhood online is a behavior that commentators made much of in the early days of the internet, particularly emphasizing the liberating potential of this realm of interaction.[71] Though more recent scholarship has provided a welcome corrective to the notion that social categories like race, gender, and class get left behind in online spaces, and to the suggestion that there is a stark separation between online and IRL, it is also true that

69. Double Union, https://www.doubleunion.org/, accessed 2/25/15.

70. Interview, Salix, New York, NY, July 17, 2012.

71. A prominent example is Turkle 1995. For critique of what is at stake in casting aside identity, see Nelson 2002: 3-4.

people like Salix have often spent time during childhood, adolescence, and young adulthood playing online games or interacting in spaces like chatrooms or IRC.[72] And they found these to be spaces for reflection about gender and sexuality, taking away the lesson that technology could be used to enhance or express true selfhood. (Salix is also an enthusiast of transhumanism, which informs a techno-utopian belief that technology can get into and enhance nuances of one's personality or body, or provide new avenues for self-expression.[73])

Salix is hardly alone in identifying an association between hacking objects or code and hacking gender. A queer-identified Belgian artist in their late thirties said in a meeting, "If you're trying to open the source code of gender, it's not right to use a closed-source operating system. . . . If you're reflecting on your gender, you will be similarly reflective about technology."[74] They suggested that for a person drawn to question their gender (and heteronormative sexuality), there is "an issue [when] the first hardware you encounter is closed, is a [Microsoft] Windows system." This shows a direct analogy in their thinking between an open-source ethos in computing and a questioning attitude toward gender; to wish to see inside and modify where one fit in a socially assigned system of gender was not very different from wanting to see inside and modify the operating system of one's computer. They also suggest that people might experience questioning and exploration from the earliest points of contact with these systems: something could feel intuitively wrong about learning to compute on a "closed" machine like a PC running Windows. The artist and the hackerspace founder above were incredibly self-aware and articulate about the parallels for them between hacking gender and hacking machines.

Similarly, in the first-person account of a trans woman in London named Sarah Constantine, a connection arises between electronics hacking and being trans:

> [As a boy growing up] I'd try to drag bits of machinery home. Old televisions, radios, mopeds, motorbikes, lawnmowers, anything electrical, mechanical, a boiler. . . . At twenty-four, I got into electronics. This is in the early to mid-Eighties. I spent three years building a huge sixty-five module synthesizer. . . . I would go to sleep thinking about electronics

72. See, for example Miller and Slater 2000; Kendall 2002; Nakamura 2007.

73. Interview, Salix, New York, NY, July 17, 2012. For more on geeks' enthusiasm for transhumanism, see Kelty 2008.

74. Fieldnotes, August 20, 2016, Montréal, Canada.

and engineering and wake up thinking about electronics and working out circuits. . . .

[Living in my assigned masculine gender,] I had created a [convincing] male persona. . . . [I] was a chameleon. But you're constantly paranoid that you're going to be found out. . . .

I just started taking hormones out of curiosity to see what it did to me. . . . When I transitioned, one of the things I found out was that about 80 percent of male-to-female [trans people] have a background in electronics and engineering.

I'm not kidding. It's so freaky. One day at the psychiatrist's—you have to see a psychiatrist for two years [to be permitted to transition medically]—there was a notice on the board in the waiting room. A self-help group had started up in someone's living room. So I went around. There are about eight sheepish-looking, ropy-looking [trans people]. And we're all sitting there in someone's living room. Things were a bit awkward. We didn't know what to talk about. Someone mentioned in passing they were into electronics. I said I was into electronics. Someone said, "I started [reading] this Ladybird book [a UK children's book series including educational titles]. It was called *Build Your Own Transistor Radio with a Plank of Wood*." I said, "I remember that book. Use the OC71 transistor." Someone else said, "Oh yes, and the OC45 for the radio frequency." I mentioned modular synths and someone said, "What chip did you use for your roger-controlled oscillator?"

"I used the Ellum 13 700."

"That's the preceding?"

"That's right, because it had linearizing diodes in the input and non-dedicated diodes."

"That's right. It came with an 11 fuel pin."

All of a sudden it's like . . . [hums the theme from *The Twilight Zone*] [television program]. What's happening? They all know the same code numbers I do? I looked at them, they were all looking back, and I realized they're all really nerdy.[75]

This is one person's account, and as such it must be regarded as anecdotal data (the "80 percent" she references is clearly hyperbolic). But it is highly suggestive, and it conforms to how others have characterized relationships

75. From an interview with Sarah Constantine (a pseudonym) by Craig Taylor 2012. See also Stone 1987.

with technologies as sites where gender identities may be transformed.[76] The high degree of interest in and pleasure from electronics' inner workings that Constantine describes is consistent with many geeks' accounts and is what sets them apart from "everyday" users of the same technologies. Her lengthy description of her own solace in electronics and the discovery that her pursuit was not singular to her stand out in her narrative. Building modular synthesizers, tinkering with ham radio transmitters or computer hardware, and coding are all practices that reflect obsessive enthusiasm for playing with and reconfiguring these machines. To reconfigure one's own gender is, for some, a logical next step for someone accustomed to "hacking on" machines and dissatisfied with or questioning their assigned gender. Constantine, the Belgian artist, and the Northwest makerspace founder each describe this, albeit in slightly different ways. Constantine describes the concrete step of "taking hormones because I was curious," while Salix describes a more abstract "questioning [of] a system and where I fit in it," wondering (half-jokingly) if this connection can be explained by "brain chemistry."[77] (Though I in no way wish to wade into debates about a biological basis for gender, in the brain or otherwise, Salix and Constantine might find much to discuss.) The Belgian artist draws a parallel between modifying the source code of a machine and being reflective about the source code of gender (and vice versa).

However evanescent some of these connections may be, there is evidence that in many tech subcultures, a high level of attention is paid to gender issues. These very much include nonbinary conceptions of gender, and people are often quite self-aware about gender issues. In the unconferences for "women in open technology," organizers had explicitly required that applicants for the events "identify as women in some way that was significant to them," leaving to applicants to determine how feminine gender applied to them. This left a good deal of room for gender presentations and

76. Trevor Pinch and Frank Trocco propose that this may have been the case with synthesizer player Wendy (formerly Walter) Carlos, who released the famous *Switched on Bach* album in 1968 (2002: 138). It is notable that synthesizers were an interest for Constantine as well. Lynn Conway, Mary Ann Horton, and Sandy Stone are other prominent historical examples of trans people devoted to electronics and computing. See Hoffman 2017 for further discussion of how trans peoples' lives may offer productive insights in critical study of technology; see also Cárdenas et al. 2012; Dunbar-Hester and Renninger 2013.

77. I do not wish to fully equate the queer hackerspace founder's account and experience of gender and sexuality in the second decade of the twenty-first century with Constantine's experience of gender transition in the 1980s. Categories of gender and sexuality are historically bound, and disciplined by, among other things, state and institutional mandates and social mores (and these are linked, of course).

identities across a spectrum, and as discussed above, much of geek femininity already traffics in nonnormative femininities. In the unconferences, these topics were discussed at length. One person who said they identified as "genderfuck" or gender-fluid also said that they were aware that "I have more street cred if I say I am a woman fighting the gender gap [in tech] than if I say I'm a gender-fluid person fighting the gender gap [so as a matter of strategy I will say I'm a woman]."[78] Sam, a queer-identified person, said, "Being in tech has changed how I identify. I didn't really care [before] but now I'm like, 'I'm a woman! Women matter! We need more women!' But if I need to throw in the word 'trans' because I think the room needs to hear that, I'll throw in the word 'trans' too because I can do that [identify that way as well]."[79] Jamie, a nonbinary queer-identified person with a very androgynous presentation, said they "felt comfortable advocating for women even if I feel [I am] particularly masculine"[80] and also said that women-only spaces were not especially comfortable, because "[I have felt] more ostracized in lesbian-only spaces."[81] Tech, as a normatively masculine domain, may seem to offer a home for people who do not identify as particularly feminine, but it is certainly not without complications.

For anyone who does not immediately feel like they fit into the normative masculinity of tech culture, reflection on gender issues is all but compulsory. Women (to say nothing of people who are genderqueer, etc.) are always needing to negotiate their identities and their presence in these communities, often feeling subtly policed in a damned-if-you-do-damned-if-you-don't way. One person said that tech itself "deprecates putting any effort into your appearance. I go back and forth. No exercise, fuck taking care of my appearance, fuck makeup, fuck plucking my eyebrows. . . . But then, [I feel like,] no, I want to take care of my appearance, because I feel good when I do this."[82] This echoes the geek stereotype of an obsessive person neglecting bodily care routines in order to hack or tinker: computer-obsessed MIT students in the 1980s were described as people who "flaunt their pimples, their pasty complexions, their knobby knees, their thin, under-developed bodies."[83] One woman agreed: "I'm not hyperfeminine normally, but when

78. Fieldnotes, July 11, 2012, Washington, DC. See Penny 2015.

79. Fieldnotes, July 11, 2012, Washington, DC.

80. See Halberstam 1998.

81. Fieldnotes, July 11, 2012, Washington, DC.

82. Fieldnotes, June 7, 2013, San Francisco, CA.

83. Turkle 1984 quoted in Eglash 2002: 49. Eglash points out that both masculinity and whiteness lurk in this description.

I'm in a hypermasculine environment, I want to be like, dude! I'm a woman. I'm not sure if this [constant self-consciousness] affects my authentic [experience of] self."[84] To this, another woman replied, "I get a comment if I wear a skirt or pants. It underscores that I'm being watched."[85] All these comments make evident that there is no neutral or default way for women* to present in tech workplaces or hobbyist spaces (which clarifies the appeal of separate spaces, as discussed in chapter 4). As I found entering the 2011 PyCon, eyes will be on you if you present in a way that deviates from the expected masculine presentation; even though this may not feel like hostile attention, it can be palpable.

Jamie (the self-described "masculine" nonbinary person, who had been assigned female at birth) expressed a poignant example of women's experiences in technical space, remarking, "I need to check my own assumptions. If I see someone [presenting in a way that is normatively] feminine, I do [initially] assume they are in marketing or HR. I don't [then] think, I'm a bad person, I [just] think [that] I absorbed some bad messages."[86] In other words, even someone like Jamie who consciously "knew better"—having been assigned feminine gender at birth, experienced technical socialization as a person who did not feel they fit into the gender binary, and gone through engineering studies at MIT—still ran up against internalized stereotypes, making knee-jerk assumptions that women were ill-suited for technical roles. In turn, women, wherever on a spectrum of gender they identify, are confronted with needing to position themselves vis-à-vis these perceptions and biases. (This can also be seen in the above discussion of women needing to choose between being perceived as likeable or competent.)

One strategy people adopted was to have a shock of vibrantly colored hair. I came to recognize this as a very common geek signifier, sported by many women (as well as some men, and many people along a gender spectrum). At a zine-making event at a feminist hackerspace in San Francisco in 2014, attended by around fifteen people, I was certainly in the minority, having uniformly bland hair in my natural color.[87] At one of the unconference events, also in San Francisco, people claimed there was strategic value in this aspect of their self-presentation: one woman with short brown hair

84. Fieldnotes, June 7, 2013, San Francisco, CA.

85. Fieldnotes, June 7, 2013, San Francisco, CA.

86. Fieldnotes, July 11, 2012, Washington, DC.

87. The colorful geek hairstyle preceded and differed from the "mermaid" pastel hair coloring trend for women that also appeared during this time period; while both possibly originated from the aesthetic of anime, the effect and signification are not the same.

containing vivid pink and green patches claimed that, "hair color is not a solution, but it's mitigating. It can draw attention, but in a way that's respectful."[88]

In other words, she felt that a bright shock of hair could provide to interlocutors a relatively neutral aspect of her appearance on which to focus their reflexive comments about her presence or appearance. This was also a strategy for genderqueer people: a tall, striking person whose presentation of self did not conform to an obvious binary gender identification said the choice to wear a waist-length, eye-poppingly purple braid was in order to have "a lightning rod for commenting about my appearance. It makes for [more] non-harass-y comments."[89] Taking the measure of these observations, it is evident that *no one* expected to *not* receive attention for their appearance. The high degree of reflection about presentation of self (including gender presentation) within diversity advocacy is therefore unsurprising. This reflection, and even self-consciousness, occurred for women (cis and trans), and for people whose identities spanned less normative spots across a spectrum of gender.

As I have suggested above, there are probably reasons specific to tech that account for the salience of nonbinary and queer identity formations in these sites. People suggested that work setups that permit telecommuting, which can include many programming jobs, can be relatively attractive options for people whose gender identities do not conform to normative ones.[90] There is also the prevalence of identity play through computing, as discussed by the makerspace founder, above.[91] And geek identity has long been bound up with certain kinds of outsider-ness, as described in chapter 2. In FLOSS and hackerspaces, geekhood carries echoes of countercultural selfhoods as well: hippies, back-to-the-landers, squatters.[92] An early account of the Burning Man festival, an annual countercultural desert bacchanal attended by many in the Silicon Valley IT industry, described it thusly:

> There are all sorts here, a living, breathing encyclopedia of subcultures: Desert survivalists, urban primitives, artists, rocketeers, hippies, Deadheads, queers, pyromaniacs, cybernauts, musicians, ranters, eco-freaks,

88. Fieldnotes, June 7, 2013, San Francisco, CA.

89. Fieldnotes, June 7, 2013, San Francisco, CA.

90. Interview, Meg, March 2012, Boston, MA.

91. Avery Dame demonstrates that online spaces were also significant as archives for queer and trans communities, which generated collective memory and served as a resource for identity formation (2016).

92. See Dunbar-Hester 2014, chapter 2; Turner 2006; Pursell 1993.

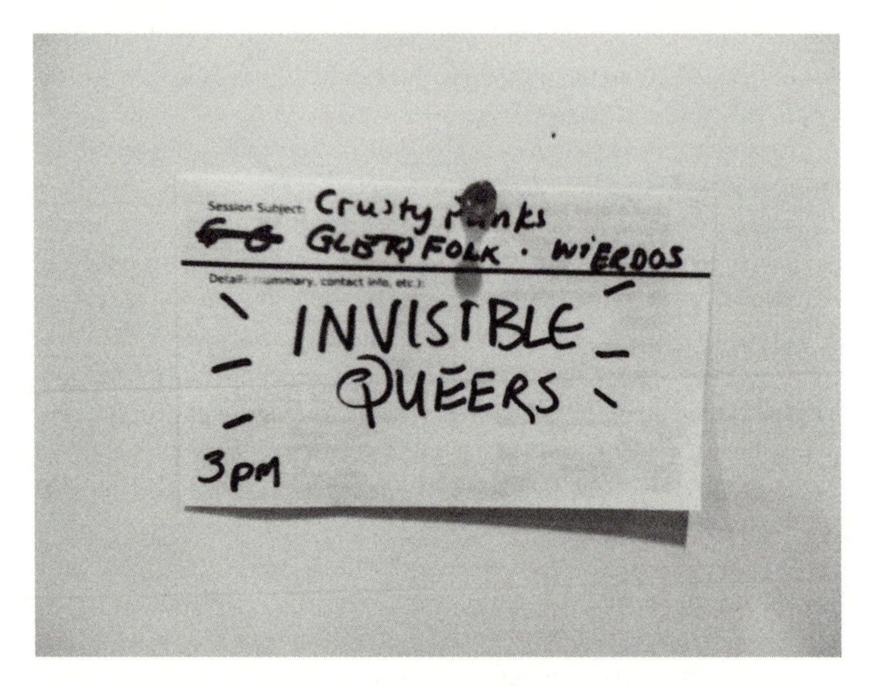

FIGURE 7.3. Handwritten note calling for "Invisible Queers" informal meeting, PyCon, March 2011, Atlanta, GA. Photograph by the author.

> acidheads, breeders, punks, gun lovers, dancers, S/M and bondage enthusiasts, nudists, refugees from the men's movement, anarchists, ravers, transgender types and New Age spiritualists.[93]

For our purposes, it is notable that members of tech subcultures ("cybernauts, rocketeers"), counterculturalists ("eco-freaks, urban primitives, nudists," etc.), and folks expressing nonnormative gender ("transgender types") were all rubbing elbows in the desert in 1995. This cultural formation is both a legacy of the Bay Area counterculture and a progenitor of present-day Silicon Valley culture, as described by Fred Turner.[94] In a direct parallel, at the 2011 PyCon, people convened a "birds of a feather" session (a spontaneous meeting not on the official conference program) for "invisible queers" (figure 7.3). *Queers* was used expansively, including not only nonnormative gender expression and sexual orientation ("GLBTQ folk"),

93. Quoted in Turner 2009: 82.

94. Turner 2006; 2009.

but also "wierdos"[95] [*sic*] and "crusty punks."[96] To "fly your freak flag high" is encouraged in communities that draw from a countercultural heritage where self-exploration and self-expression are valued. This also fits with a mythos that surrounds programming cultures, which have a long-standing commitment to the idea that your personal identity is irrelevant if you can make valued contributions to a technical project. (We should not take this mythos at face value, of course, as programming has also been associated with sexism and heteronormativity,[97] but it does constitute a resource for justifying nonnormative personal expression in the context of FLOSS.)

Having established the varying ways that diversity advocates understood and grappled with their mission to expand gender diversity—and to promote an expanded notion of gender diversity—in open-technology communities, I now turn to their diversity mandate as it pertains to race and ethnicity.

Positionings: Ethnicity, Race, Nation

As noted throughout this book, gender diversity (often meaning "more women") was the most consistent formulation of diversity in open technology communities. This does not mean that it was the only aspect of diversity to which advocates attended. In fact, people advocating for attention to racial and ethnic diversity pushed back on the *diversity as gender* framing. One person who cofounded a "people of color-led" makerspace in Oakland, CA, wrote in 2014 on a social media platform, "Diversity means more than gender. Blacks and latin@s are conspicuously underrepresented in tech, too. That's why we started [our people of color makerspace]."[98] Gender diversity was very important for the founders of this space too; they were, however, making an explicit effort to situate themselves in contrast to dominant hacker or geek identity in which whiteness has historically been hegemonic.

As noted above, whiteness has been a dominant though often unmarked identity category in the cultural milieu of hacking. The celebrated boy-hero with dazzling mastery over technology has been a media trope since the early twentieth century. Radio historian Susan Douglas writes, "[The public]

95. See Patterson 2014 on "weirdos" in tech.

96. For more on punk subculture identities, see Portwood-Stacer 2013.

97. Hicks 2017.

98. Twitter, June 28, 2014.

did not know they were surrounded by an invisible and mystical realm to which youthful wizards such as [radio hams] were privately gaining access."[99] As this statement suggests, there is striking continuity in how radio hams and more recent hackers have been represented, primarily as young men with great technical aptitude whose powers awe the public. For both hams and hackers, though, the public's attitude of dazzlement could shade into ambivalence; at times, they have been subject to suspicion over their motives and their power to not only do public good but public harm with their technical prowess. Hams pushed back on government efforts to curtail their communicative freedom during World War II, citing the importance of the ham communication network to a public safety mission, and presenting ham tinkering and communication as a "resolutely apolitical" activity.[100] Modern-day hackers like Julian Assange or Edward Snowden are familiar with public ambivalence toward their activities as well.

The "electrical priesthood" of white men was elevated in relation to and at the expense of women, rural and lower-class people, African Americans, indigenous people, colonial subjects, and immigrants, as documented by cultural historian Carolyn Marvin.[101] For white youth, beginning in the early twentieth century, electronics tinkering was understood as a path to gainful employment in white-collar technical occupations. A typical instance of this cultural association can be seen in a 1955 *Popular Electronics* cover; we see a young white man wearing a literal white collar while operating ham radio apparatus (figure 7.4). He is almost certainly located in a "ham shack," a basement, attic, or garage where men and boys carved out a masculine space from the feminine domain of the home, as documented by historian of technology Kristen Haring.[102] There is notable continuity between this image and the early days of "home-brew computing" as illustrated in this image from the 1970s, in which a young man has built a programmable minicomputer in his garage (figure 7.5).

Though these leisure activities were a proving ground for whites' entry into technical occupations, by contrast historian of technology Rayvon Fouché has documented how the meaning of tinkering and inventing for African Americans was circumscribed by racism. Blacks were largely denied access to both shop culture and school culture,[103] two avenues where white

99. Douglas 1987: 188.

100. Haring 2006: 156.

101. Marvin 1988.

102. Haring 2006, chapter 6.

103. See Slaton 2010.

FIGURE 7.4. *Popular Electronics* magazine cover, February 1955.

Americans honed their technical skills through informal apprenticeship or formal education and found routes into scientific and industrial cultures where credentialed invention occurred.[104] Furthermore, to reprise historical background from chapter 2, acts of black invention could be framed in

104. Fouché 2003: 12–13.

FIGURE 7.5. Homebrew computing, ca.1975. For more on the Homebrew Computer Club, see Turner 2006: 115–16. Courtesy of Bob Lash, http://www.bambi.net/bob/homebrew.html.

a negative context: a technological innovation that might be celebrated as an act of "ingenuity" in a white inventor might be received by whites as an act of "laziness" or an effort to sustain poor work habits if advanced by a black inventor.[105]

Thus, returning to the topic of ambivalence toward hackers or hams whose activities might at times be met with suspicion, it is apparent that whiteness has been a resource for white technologists, understood as non-threatening, "good" actors in society, dazzling wizards instead of criminal threats or layabouts. Electronics tinkerers who were white could cast even their mischief (as practiced by not only stereotypical teenage hackers[106] but also, for instance, radio hams who "trolled" the US Navy with obscene messages in the 1910s[107]) as a stepping stone to cultural legitimacy in the form of high-status white collar employment. Probably needless to say, such behavior would likely be met with harsher social or even legal sanctions for

105. Fouché 2003: 12. Fouché quotes a white lawyer weighing in on a dispute over a black inventor's patent application as saying, "It is a well-known fact that horse hay rake was invented by a *lazy negro* who had a big hay field to rake and didn't want to do it by hand" (2003: 12; emphasis Fouché's).

106. Schulte 2013, chapter 1.

107. Douglas 1987, chapter 6.

members of social groups whose expressions of dominance (technical or otherwise) would not be celebrated by the wider society.[108] Black inventors might not strive for subversion as much as for legitimacy; some were motivated by the goal of acquiring proper credentials to assimilate into white society.[109] Upwardly mobile middle-class whites, on the other hand, did not need to assimilate, and their acts of pranking can be understood as an expression of not only playful engagement with technology but also social dominance (though neither hams nor hackers as a class would necessarily acknowledge this consciously).

This historical background illustrates why, as Ron Eglash has argued, geek identity limits the capacity of people of color to take up its mantle whole cloth without critiquing it or innovating upon it.[110] Because of this legacy, the people-of-color-led makerspace in Oakland self-consciously called itself a *maker*space, not a *hacker*space. June, a founder who is East Asian-American, said in an interview, " 'Makerspace' is more welcoming than 'hackerspace.' We do a variety of activities here, and we want people to be attracted to us, not to put on [events] where we tell them what to do—we want there to be cross-pollination [between what they are already doing and our mission]."[111] June's comments are revealing. She states outright that the "hacker" label and identity would likely not "attract" many people of color. Though she did not go into detail, the above discussion of hackers as people who enjoy technical problem-solving and are often willing to flirt with mechanical and electronic mischief—such as lock-picking or phone phreaking, as well as probing computer networks where one is not authorized to be[112]—suggests that members of social groups who have historically been assumed to be criminally suspect, and have been surveilled to greater degrees than middle-class whites,[113] might be hesitant to embrace

108. In our present day, Arab American or Muslim hackers might be met with extreme suspicion. The 2015 case of Ahmed Mohamed, a Sudanese American high school student in Texas who built a homemade clock and was accused by his teacher of making a bomb, brought forth charges of ethnic profiling and Islamophobia (Chappell 2015).

109. Fouché 2003: 6. See also Nguyen 2016 for a discussion of hacking as breaking *into* technocultures from which certain groups have been excluded, as opposed to breaking *out* of sociotechnical limitations (Nguyen's emphasis).

110. Technological curiosity or ingenuity in African Americans would not be welcomed or rewarded by the wider society, which, Eglash argues, led them to innovate Afrofuturism as a black cultural strategy for engaging with technology (2002).

111. Interview, July 24, 2014. For a related critique of "maker," see Chachra 2015.

112. See Levy 1984; Coleman 2012b.

113. Eubanks 2012, chapter 5.

the "hacker" identity. Perez, who self-identifies as a "bi-cultural Latina," said in an interview that she found the stereotypical hacker aesthetic to be off-putting. As someone who spent a lot of her time in the infosec community (digital privacy and security), she said that the "pressure to have to wear all black and be in an intense vampire movie, [have an] emo or goth [aesthetic]" was, to her, "cultural imperialism, too Anglo, too German. . . . We don't have to all be hackers in that stereotypical way."[114] Though she presented this topic half-jokingly, she is serious in her critique that the dominant aesthetics reflected cultural positioning that was literally foreign to her, and which she had to learn how to navigate upon entering tech workspaces and hobbyist spaces. (She said that over the years she had modulated both her femininity and her ethnicity in order to be accepted in tech circles; she claimed that she became more "tomboyish" because she didn't feel she could express the normative femininity of her cultures of origin in open source, echoing some of the above discussions about gender.)

June, the Oakland makerspace founder, also stated, "there is almost no support for blacks and Latinas—being people of color [in tech] is a different challenge."[115] This comment suggests a number of issues surrounding race and ethnicity in open technology cultures. First, consciousness regarding gender diversity does not necessarily translate to matters of race. She alludes to "challenge" as being part of the allure of tech work—coding, solving problems—but she states openly that for people of color, not all of the challenges are necessarily of a positive or motivating sort. She indicates that, as noted above, whiteness is the most prevalent and often unmarked category of racial identity in open technology, which "others" must position themselves in relation to. And within these "other" categories, there are nuances and gradations that make for a multitude of experiences surrounding race, ethnicity, and belonging in open technology communities.

June discussed her own background as an East Asian American. She said, "Being Asian is interesting because you have some degree of privilege, but [in reality] a lot of Asians aren't privileged."[116] As American studies scholar Lisa Nakamura has explored, Asians are in a special position in terms of assumed techno-proficiency; this assumed special position elevates certain Asians and

114. Interview, Perez, November 4, 2016.

115. Interview, July 24, 2014.

116. Interview, June, July 24, 2014. In June's words, "Asian American is an invented category," highlighting the fact that the experiences of Chinese, Taiwanese, Japanese, Cambodian, Hmong, Vietnamese, and other immigrant groups in the United States are not necessarily similar, even though they all tick the same box on a US census form.

Asian Americans symbolically while rendering less privileged Asians and Asian Americans invisible.[117] She writes, "The ongoing cultural association with Asians and Asian Americans and high technologies is part of a vibrant and long-standing mythology."[118] Multiple American interviewees with South Asian and East Asian ethnic backgrounds described ambivalence about their experiences with race in technical settings including education, hobbyist or volunteer spaces, and workplaces. Anika, an American woman in her thirties whose ethnic background is South Asian, said that she was sure that when she first tried to learn to program in a college course, she was not subject to as much challenge or pushback from peers as she might otherwise have been. She said, "To be frank, some sort of racism helped me—'Indians are good at coding' stereotype—people looked at me as Indian, not as a woman."[119] She opined that she did not have negative experiences with peers questioning her competence because of racial stereotyping, even when she was a novice and not particularly good at coding. She said did not enjoy her first foray into learning to code; she had enrolled in the course in part due to family pressure to study science or engineering and said that when she did not like the course or do well in it, she "used that as an excuse" to evade that pressure. Crucially, though, she did not feel out of place in the environment because of gender or race, and she later returned to learning to code, driven by her own curiosity and being a "power user" of FLOSS applications.[120] It is perhaps relevant, too, that Anika's presentation of self was not especially feminine: when I met her over the course of several events, she sported short-cropped hair, no makeup, a t-shirt, and khakis or jeans. I do not know what her appearance was like years earlier when she was initially learning to code, but it is possible that being seen as Indian first and woman second was a reaction she elicited in part due to how she presented herself. As discussed above, many women* in FLOSS do not present as normatively feminine, and they describe receiving different reactions to themselves based on whether they were presenting as "more femme or more butch," were made up or not, or were wearing a t-shirt and jeans versus something more formal or more feminine.[121]

117. Nakamura 2007, chapter 5.

118. Nakamura 2007: 180. For more on the stereotype of Asians as "good at coding," see Chun 2011.

119. Interview, Anika, July 7, 2015, New York, NY. See Amrute 2014: 115 for a discussion of IT workers' ambivalence toward the "good Indian coder" stereotype.

120. Interview, Anika, July 7, 2015, New York, NY.

121. Fieldnotes, June 8, 2013, San Francisco, CA.

Paul, a man in his late twenties who was active for several years in diversity efforts in the Python community, was also of South Asian descent (his family was from India and he had been raised in the UK and US). He helped launch volunteer-run programs to teach "women and their friends" Python in Boston that were replicated in many cities. He reflected on why there were not more South Asians in FLOSS, which he viewed as especially curious given that they were well represented in many technical fields, and that Indians' English proficiency is high. (Though FLOSS contributors are based all over the world, English is a dominant language in FLOSS.) Paul said that he himself had been introduced to FLOSS in his youth by a "white kid" at summer camp, and then had pursued this burgeoning interest on his own. But he said that "if you're an immigrant from South Asia like I am, in your conversations with your parents, it's unlikely that you would be encouraged to [just] fiddle with a computer."[122] In other words, while he (like the other South Asian-American respondent above) was urged to pursue science and engineering by his parents, the "fiddling," tinkering, or self-guided exploration—about which nascent hackers or FLOSS geeks are enthusiastic (and which becomes second nature for them)—was not something in his repertoire based on his cultural background. While he did eventually cultivate a fusion of joy, curiosity, and agency in his relationships with computers, this attitude was something to which he was first exposed by people whose cultural background was different from his own. The author of a blog post about South Asians in FLOSS (using Debian as his example) writes that, "My mother reports that schools in India focus on memorization instead of creativity. This can leave little room for extracurricular pursuits [or self-guided exploration]."[123] Just exploring, and less practical interactions with computers—such as "finding out about features even if I'm not using them"—were experiences Paul and others had to pursue without family encouragement.[124]

Along similar lines, Helena, a self-identified mixed-race woman in her thirties, offered insights about the cultural assumptions of many people in open technology circles, including some of those fervently advocating for diversity. Based in the San Francisco Bay Area, she had witnessed over the years a number of initiatives to promote diversity, both in industry and outside of it. These included the birth of hackerspaces and makerspaces centered

122. Interview, Paul, March 2012.

123. Laroia 2009.

124. Interview, Paul, March 2012.

on women* and people of color, and the unconferences for women* in open technology that I discuss in this book. In an interview, Helena said that her own family background had been precarious in multiple ways; her parental support was unstable, and any economic security was hers alone to achieve. This meant that unlike a lot of other professional colleagues in tech, she did not feel like she had a safety net to fall back on if a venture failed, and she had to contribute all of her own savings and retirement earnings without the expectation of inheritance or family support through an employment lull (leading her to prioritize paid work over voluntaristic enterprises, which she felt set her apart from peers). Living in famously expensive San Francisco, she speculated that colleagues and friends who could afford a lot of time for voluntaristic projects were accustomed to more financial security than she was. She also had not had the opportunity in childhood and young adulthood to explore computing as an affective pursuit; by the time she discovered and fell in love with the "analytical mindset" of coding, she was around twenty years old. Interestingly, she did not feel that she was at a disadvantage professionally per se, but did indeed feel that way in FLOSS. In some ways, she echoed the statements of the South Asian and South Asian American people I quote above whose family backgrounds had not encouraged the same pursuit of the affective dimensions of computing that they observed many white youths' families had.

And these factors seemed to set Helena apart from many of the other people I met in the course of this research. Her attitude toward many of the diversity advocates she encountered was, if not jaundiced, at least guarded. She questioned the sincerity of people advocating for women without doing the also-challenging work of intersectionality.[125] She said, "How many [people in] diversity roles are white women? . . . White women are not incentivized to advocate for others because they feel what they have is precarious [relative to white men]."[126] In other words, she wondered if white women, having struggled for standing and visibility in tech projects, were then inclined to consolidate the power they had, as opposed to continuing to advocate for a greater spectrum of diversity that included people unlike themselves, some of whom might challenge them. She felt that her own ability to find common cause with many (though not all) white women had been hindered by dynamics where white women did not really want to advocate

125. Crenshaw 1991.

126. Interview, Helena, August 14, 2016.

for change beyond improvement in their own stations.[127] Helena added that she felt that white women were gatekeepers between women of color and projects and groups trying to do diversity better, because white women, as women, could easily stand in for movement toward diversity. She said that her experience had been, "If you raise anything [critical] as a woman of color, you will get backhanded backlash, *House of Cards* shit [a television program in which characters engage in political scheming and double-crossing], [they will] not invite you to things, or make up or misconstrue things [about your intent]. It's really hard unless you have a white female ally."[128]

Writing on an electronic mailing list, a woman in India made an explicit critique of what she perceived as lack of solidarity from white women, especially those in the Global North. The incident in question is not especially important.[129] Of greater significance is how her words invoke race and reflect differences in global positionings among diversity advocates:

> While I appreciate the work being done by American feminists, the focus is entirely on the needs of white American women and the needs and experiences of colored women, especially those women from other nations are completely ignored and negated. . . . When men do it to women, [it's] called sexism, *so what is the term when white American women silence colored women*?[130]

These statements directly restate some of Helena's claims, though adding the dimension of global positioning.[131] She questioned whether white women in the Global North were allies for women of color, especially those in the

127. Reflecting on these dynamics from the opposite side, as it were, Meg, who is white, observed that race as an axis of advocacy had been relatively muted for some people in part because, "White women who were there [doing diversity advocacy] . . . didn't think they could speak for others" (interview, March 2012, Boston, MA). Further underscoring frictions, some women of color felt they had been tokenized by white women diversity advocates (personal correspondence with author, March 7, 2019).

128. Interview, Helena, August 14, 2016.

129. The conflict concerned a rehashing of how some people had been banned, in her view unfairly, from an electronic mailing list, which emerged as a topic on a different list in response to a different incident. See chapter 3 for more on the prevalence and significance of these backchannel communications.

130. Email,—to [Women* list], July 31, 2015. Emphasis added.

131. Studies in different national contexts reveal, unsurprisingly, quite different coconstructions of ethnicity, gender, and technical identity. See Kuriyan and Kitner 2009 on India and Chile; Lagesen 2008 and Mellström 2009 on Malaysia; Lindtner (forthcoming) on China; and Kim et al. 2018 on South Korea; for comparative work across the Global North, see Abbate 2012; Hicks 2017; Corneliussen 2012.

Global South, given that it was easier to improve their own positions (vis-à-vis white men and people in the Global North) than to have solidarity with people like herself.

The point of relaying these sentiments is not to adjudicate what "really" happened, nor to cast blame. We should understand these words as a sincere expression of pain and frustration that can arise for people hoping to build coalitions across difference, especially those who feel more marginalized or multiply marginalized. As Helena, the mixed-race woman in San Francisco, observed, "[It's] dangerous to be the messenger [of a critical message about a project, or about a diversity advocacy intervention] if you're a woman of color. They [we] need to create pathways through white women allies . . . [and then] *whoops!!* [if you are not careful, you could set back the issue you're working on, or find yourself ostracized]."[132] A deep commitment to empathy and trust was, she felt, lacking in many of the diversity advocates she encountered. Her rebuke of some white women diversity advocates presents an occasion to revisit the slide introduced in chapter 3. Here, whiteness is an unmarked category, often unspoken yet looming large in many FLOSS and hackerspace sites (figure 7.6). An image like this reminds us that diversity advocates, who have often begun with highlighting gender, are working through multiple thorny issues of identity and difference.

We do not need to know more of the incidents to which Helena alludes in order to understand that these dynamics can have a deleterious effect on open technology projects where diversity advocacy is occurring. The perceived difficulties of forming coalitions and working across difference have the potential to damage or undermine diversity work, which requires trust. These problems are perhaps especially acute in projects that run on good will and are organized around a voluntaristic ethos where much energy is generated through personal relationships.[133] This is not a reason to simply despair of diversity work, but it does indicate that there is ample room for additional consideration of how difference—across lines of race, gender, language, class, (dis)ability, nationality, and so on—can be brought into generative exchange within the shared and multiple imaginaries of power and participation in open technology. This section has illustrated that, though their experiences vary, many people of color, positioned in multiple ways, feel that they have to grapple with hegemonic whiteness or their own "otherness" in open technology communities. Even for people

132. Interview, August 14, 2016.

133. Polletta 2002: chapter 6.

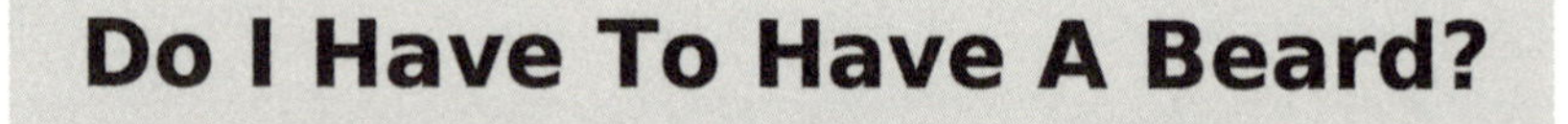

FIGURE 7.6. "Do I Have to Have a Beard?" slide. Though a humorous and effective way to question a default assumption that FLOSS participants are men, this slide inadvertently reinforces whiteness in FLOSS. Wallach 2010. Courtesy of Hanna Wallach.

who are relatively advantaged or those who experience "helpful racism," there is a perception that difference must be negotiated.

Conclusions: Social Identity, Power, and Expert Knowledge

The above discussions illustrate many (though certainly not all[134]) of the important contours of discussions about people's positioning in open-technology cultures occurring within diversity advocacy.[135] They are

134. Though it is outside of the scope of this account, there would be a longer story to tell about diversity advocacy in FLOSS around disability and ableism. See for example Ellcessor's discussion of Dreamwidth (2016); Forlano on hacking the feminist disabled body (2016); Pullin on open-source prostheses (2011: 226); Salehi and Bernstein on Mozilla and accessibility (2018).

135. It is important to recall that these conversations are all located in North America and Europe from the 2000s onward; the conceptions and conditions of race and gender that are examined here are situated in time and space.

representative of a pivotal moment wherein open-technology cultures have moved significantly toward consideration of categories of social identity and social difference as they reflect upon their own constitution; they reflect an emergent politics of recognition.[136] Doubtless, this is in part because FLOSS cultures have over time matured, witnessing the relative institutionalization of many projects and tinkering with their cultures along the way.

These conversations reflect advocates' admirable impulses to respond within their own subcultures to wider social problems of social inequity and segregation.[137] And yet, they are almost immediately caught on the horns of multiple dilemmas about how to engage categories of social difference as they relate to open-technology communities. Ironically, some of the most nuanced ways of addressing social difference are rather at odds with realistic goals for voluntaristic groups brought together by the shared social imaginary of open technology. For example, a white man in his thirties who had founded a Python-based project renowned by diversity advocates for being hospitable and welcoming, said in an interview, "African Americans are underrepresented [in open technology], but more than half of Americans are women. If we do [outreach] efforts focused on women, a lot of the women we've been attracting aren't white, [this is] increasing the spectrum already. [Racial diversity] won't completely take care of itself but [outreach to women] will address most levels."[138] He is correct that African Americans are underrepresented in open source.[139] And he is also obviously correct that not all women are white, and that outreach to women may signal to members of various groups that there is attention being paid to inclusion. At the same time, as he acknowledges, to lump racial inclusion in with gender inclusion is to paper over some of the unique features of cultural and personal histories

136. Markell 2003.

137. danah boyd claims that self-segregation in the United States is increasing due to not only online self-selection into homophilic social groups but structural factors such as increasing school choice and the privatization of the military; in her conception, social media amplifies segregation but does not cause it (2017).

138. Interview, July 3, 2012. See Poster 2008 for a discussion of some of these dynamics in the (transnational) corporate sector.

139. Callahan et al. broke "representation" down into *community member* versus *contributor* and found that just looking at "community members" overcounted the African Americans and Hispanics in the FLOSS project they studied; when looking at the ethnicities of *contributors* to the project, the presence of African Americans and Hispanics was even more limited (2016). Again, I use *Hispanic* because that is the language of this report. Of course, it is not possible to generalize their findings about one project to FLOSS as a whole, but their finding is provocative nonetheless.

that people from different racial and ethnic groups have experienced vis-à-vis technological cultures and online cultures.[140]

In particular, the experience of *vulnerability* is one that stands out in the accounts of women of color in open technology. In describing a hateful instance of harassment to which she was subjected by online attackers, Helena, the mixed-race woman quoted above, said that she felt her experience was intensified as a woman of color. She said, "I was way more scared than women not of color [would be]."[141] Though of course fear is a subjective experience, that is beside the point. In saying she felt "more scared," she is referring to how misogynistic attacks online are also racialized. In addition, her more mundane experiences in open-technology communities—as opposed to the one where she was viciously attacked—left her unsure who to trust. Like the woman from India, she did not always feel that white women were her allies. This points to a major double bind for advocates collapsing *diversity* into *gender* or *women*. It leaves them poorly positioned to respond to how *differential* social statuses and vulnerabilities play out.

While these issues are important, so far they largely mirror the social experience of racial and ethnic minorities in the wider culture and online, where whiteness is hegemonic. They are not specific to open-technology communities in particular. One issue that stands out for further consideration is the fact that open-technology communities are *voluntaristic*. Unlike institutions of higher education or workplace relations between employers and workers, whose terms of association are formally overseen by regulation and the courts, these sites are predicated on elective associations between members. As cultural historian Thomas Streeter writes, in "creat[ing] social and organizational structures that by their design could motivate individuals to collaborate," engineers whose practices laid the groundwork for open source predicated the existence of these communities on affective commitments and the motivation to collaborate.[142] While inclusivity and expressions of good will toward others are laudable as values and as practices, the terms of association in these communities are less enforceable than in more formal institutional relationships. And these elective associations are

140. The group Black Girls Code is a not-for-profit organization that focuses on providing technology education for African American girls, founded by electrical engineer Kimberly Bryant in 2011 "to provide a more comfortable space for black girls to learn computing, *one in which being black and female would be the norm rather than the exception*" (Abbate 2018: S151, emphasis added).

141. Interview, August 14, 2016.

142. Streeter 2010: 105.

especially vulnerable to strain and even dissolution when personal relationships are strained.[143]

Another distinct matter is the relationship of these communities to technical expertise. It is incredibly difficult to conduct technical decision-making in a radically democratic manner. Technical decisions create path dependencies, both materially (within the artifact) and in terms of the reputational capital or acclamation of the people who recommend and shepherd along a given technical solution. Open source has traditionally solved this by a reliance on rough consensus and running code, combined with a mythic ethos of meritocracy. As Gabriella Coleman writes about decision-making in Debian, "there is discomfort with the idea that technical guardians could . . . exercise their authority without consulting the project as a whole, thereby foreclosing precisely the neutral, technical debate that allowed them to gain their authority in the first place."[144] In other words, there is a profound tension between a democratic ethos whereby "technical guardians" ought not make decisions without consulting the wider base of project members, and the fact that the technical guardians have been anointed precisely for their expertise.

Though the example of Debian shows that open-source cultures have grappled with these tensions and worked out ways to resolve them, the addition of diversity advocacy around categories of social identity is a new frontier, and a challenging one for open-technology communities. As noted above, technology has been a cultural site where gender has been constructed and maintained over time; it has also been a site where white dominance has been constructed and maintained. These conditions predate and exceed the contours of the open-technology communities where advocates are currently wrestling with diversity and inclusion. As technical expertise and decision-making becomes a domain where people reframe and contest the notion that technological production is guided by either neutral values or simply an affinity for the best technical solutions, new possibilities are opened up for democratizing technical projects.[145] At the same time, meaningful interventions into technical cultures must take into account both the history and present of inequitable social conditions surrounding the

143. See Polletta 2002, chapter 6, for a discussion of friendship as a basis for political organizing in the women's liberation movement.

144. Coleman 2012a: 127. Coleman argues that Debian members reconcile chronic anxieties about trust and authority vested in project leaders through "cabal humor"; the ultimate cabal joke is "there is no cabal" (ibid: 127).

145. Harding 2016.

creation of expertise and the arcane, esoteric nature of expert knowledge itself. In the open-technology communities where these conversations are happening, it is unrealistic (and probably undesirable) for discussions of social identity not to percolate up. And yet reconciling the expert nature of expertise, and the lived experience of some participants that "open source is so fucked up if you're a person of color, or a woman, or a person with disability. . . . It's not that women of color don't want to be involved, it's that they're being cockblocked,"[146] as an interviewee vividly put it, continues to be challenging for voluntaristic communities, to put it mildly. In combination, a democratic ethos, technical expertise, and unequal social identities make for a potentially bewildering set of encumbrances for voluntaristic communities to bear. It is exceedingly hard to dismantle white hegemony, and some of the more realistic goals for voluntaristic groups, such as reaching out to women, are inadequate to counter deeper social issues of social inequity and segregation.

One final matter warrants scrutiny. Wrestling with social identity and inclusion, many diversity advocates employ a vocabulary of *representation.* This can be seen in the above statement, "Diversity means more than gender. Blacks and latin@s are conspicuously underrepresented in tech, too," or in this utterly typical sentiment posted to an electronic mailing list, "In the fields of technology in general and hacking in particular cis and trans* women and queer folk are represented in reduced numbers."[147] The language of representation is an easy one to latch onto. It is ubiquitous, and its intent is, to a large degree, salutary: the people who use it are attempting to usher in positive social change and more equal conditions in their communities.

However, it is worth interrogating representation and its potential deficiencies. In the words of sociologist of culture Herman Gray, "abstract notions of rights and freedom and their expansion to new subjects elide the social salience of race and gender as a basis of inequality even as it culturally recognizes and celebrates differences."[148] Though Gray's critique is directed primarily at media representation, it offers important lessons for open-technology cultures as well. FLOSS in particular has been a fertile cultural site for the articulation of liberal ideals, predicated on an expansive

146. Interview, Helena, August 14, 2016. "Cockblocked" is vulgar slang (centering on male sexuality) for an action that prevents someone from having sex, or, metaphorically, that thwarts someone's intentions. As noted above, without contradicting Helena's statement, disability is something of a special case in FLOSS.

147. Email—to [Hacking List], June 25, 2015.

148. Gray 2013: 466.

notion of individual sovereignty.[149] Though open-technology cultures are also involved in deeply collective forms of imagining and collaboration—and diversity advocacy is good evidence of this—their commitments to liberalism can make it especially challenging for them to grapple with the historical reality that race and gender *form a basis of social inequality* that is not easily overcome by adding new sorts of people (whose differences are "recognized" and "celebrated") to the same cultural logics.

The quest for representation is also limited by how it is situated within "the context of market sovereignty and consumer choice organized by a logic of difference."[150] Throughout this book, I argue that open-technology cultures' interest in diversity is not defined in a singular way. This is not a problem in itself, of course. But there are potentially profound tensions between those who seek to expand open-technology participation on grounds of social justice, which includes many diversity advocates, versus those who seek to expand market potentials (commonly framed as "diverse developers will result in serving or capturing a diverse consumer market," a discourse strongly emanating from corporations in particular). In the case of people whose social identities have historically been regarded as "less than," to achieve recognition as a market is not a fully unalloyed good, as it may come at the expense of mounting political challenges to mainstream culture.[151] Though subjugated social groups generally have good reason to prefer visibility and representation over absence, there is ample reason to believe that various forms of representation do not unambiguously amplify political potential.[152] Put differently, though many diversity advocates in open technology aspire to a higher social good than the cultivation of consuming or producing subjects, they are slipstreaming on cultural currents that are much more powerful than they are. Many of these currents are more devoted to producing subjects and goods for markets than to enhancing social justice.

To frame diversity advocacy as a politics of representation is a limitation if an underlying purpose of open-technology cultures is to do something with a different political valence than representation can capture. It is not desirable for voluntaristic groups formed around the affective pleasures of

149. Coleman 2012a.

150. Gray 2013: 465.

151. Sender 2005.

152. Strathern critiques representation along related lines, noting that it essentializes people by sorting them into membership groups. She adds that it also raises another problem: the matter of which individuals are entitled to represent "their" groups (2005).

technical engagement to succumb—in their quest to promote equality—to a hierarchy of suffering, which is one potential pitfall of elevating representation as a concern above all others. Nor is it reasonable to expect these groups to solve social problems like segregation and separation encountered within spaces like schools, neighborhoods, and civic institutions.[153] Recognition within a segment of culture affectively devoted to technological production is not equivalent to just conditions within the wider society, and while the former may be ameliorative, it is no substitute for the latter.[154] Lastly, because social identities are "intrinsically incomplete and open-ended," in the words of Brian Wynne, to latch onto *fixed* social identities as a main plank of intervention is potentially quite problematic.[155]

To push past representation as a goal is perhaps to tap into more political potential in coalition across difference (without seeking to mask difference).[156] The cultivation of self-study and habits of fairness are worthwhile endeavors for voluntaristic groups of technologists. If there are politics or goals to be identified that exceed market cultivation, let them be named, and not be conflated with representation. In other words, what *is* the basis for the shared social imaginary that brings people together around open technology? How does thinking about diversity, and advocating for it, potentially transform open-technology projects and subcultures? In framing concerns about the constitution of open-technology cultures as relating to social identity and social position—including race, class, ethnicity, gender, nation, disability status, and sexuality—what is gained? And what has yet to be captured in advancing or reframing the animating impulses of open-technology communities? Might redrawing concerns somewhat away from the *representation* of various groups, and closer to coalitional production of not just technologies but democratic social relations, alter how political energies can be discharged? Lastly, to what extent can these issues be understood as challenges to be tackled within voluntaristic technological production, and to what extent do they so exceed it that the frame has to be wider? This chapter has ended with questions, not answers. At the very least, it has highlighted a central paradox for diversity advocates hoping to transform the wider culture through a focus on representation in open-technology cultures.

153. boyd 2017.

154. Markell 2003.

155. Wynne 1996: 42; see also Gilroy 2019; Markell 2003.

156. Arvin et al. 2013; Collins 2000; Haraway 1991a; Markell 2003; Prasad 2016.

8

Conclusion: Overcoming Diversity

> We're just doing workshops [for women in Python], but we're not sure of the goal or outcome. It's not really [job training], because there's still a big gap [between what we teach and what people would need to know to get hired]. . . . I actually don't really know why we're doing it. We haven't been reflecting.[1]
>
> —ELISE, WASHINGTON, DC, 2012

Something critical is at stake concerning how people interact with computing technologies, and who people are when they interact with or use computers. As cultural historian Thomas Streeter has argued, computer use is a site of selfhood construction, occurring within a history and set of social relations surrounding networked computing. When people use computers, they are often enacting simultaneous modes of selfhood whereby they are controlling something, reaffirming agency, and even hoping for self-transformation.[2] Some of this occurs vis-à-vis a business or work self and some as a personal self, but these are often blurred. The use of networked computing is often mundane in practice though suffused with great moral import. When diversity advocates attempt to open up computing cultures—especially to make good on the promise of openness of open technology for more people—they are trafficking in these interpretations of computing,

1. Fieldnotes, July 2012, Washington, DC.
2. Streeter 2010.

hoping to distribute the agentic and self-transforming relationships they believe they have with computers more widely (for this is part of what *openness* implies for them).

Elise, a thirty-something programmer and mother of a toddler who led workshops in Philadelphia for women to learn Python, made the above remark at the unconference for "women in open technology." As she indicates, while there has been momentum in recent years towards addressing unequal rates of participation in open technology, it is somewhat less obvious what is driving these initiatives. As she notes, part of her group's impulse had to do with cultivating workplace preparedness, especially in the wake of the financial crisis that began in 2007. But at the same time, she readily admits, these workshops could not stand in for job training per se, and that was never her or other diversity advocates' main intent. Diversity advocacy felt sufficiently urgent as a site to discharge inchoate political energies, but she was not certain of the goal or outcome of her efforts.

This book has chronicled efforts like Elise's Python workshops—grassroots efforts within open-technology communities to reconfigure these communities. Though she claims that "we haven't been reflecting," there is far too much activity and discussion here to take that claim fully at face value. Diversity advocacy is an often a prefigurative but meaningful expression of critical agency for its proponents. At stake for members of this social formation are large, important, and often quite abstract social goods, including democratic participation, agency over technology, and often social justice; all of these concerns are imbricated for them. The preceding chapters explored a variety of diversity advocates' interventions, which include changing rules and norms in open-technology communities, creating separate spaces for feminist hacking, bringing to the surface other political concerns like militarism and colonialism, and questioning the makeup of open-technology communities. This conclusion considers not only the consequences of these interventions, but the framings that undergird them. The current raft of initiatives—and parallel critiques about representation and equal treatment that have suddenly emerged in a number of other spaces in the broader culture, à la the "#MeToo movement" or #OscarsSoWhite—make for an opportune moment to draw out reflection and analysis that for Elise have been too fleeting.

On balance, it is worthwhile that these efforts have begun to open up conversations and take conscious steps toward transforming open-technology communities into more inclusive environments. This is undoubtedly valuable for those who are already in these communities but have felt ambivalent about some of the cultural aspects that they perceive as injurious. It is certainly also useful for some people who would be there had they not been turned off. During the period of this research, many of these communities have initiated internal conversations, set new norms, and cleaned house in ways that probably do improve their communities for people who were already there or on the margins and just needed a little more support or a little less discomfort to join.

Another important critique, raised by feminist hackers, is the degree to which engagement with open technology is or is not meant to be about *technical artifacts per se*. Many diversity advocates advocate for reframing hacking itself as *primarily about social relations*. For them, this opens up the capacity to consider who is and isn't set inside the frame of this activity. Hacking has greater significance in terms of association among its members—the relationships between people—and is less preoccupied with outcomes; it is moreover ambivalent about valorizing technology itself. Building social relations and infrastructures of care are paramount.

All this being said, much critique that peeks through diversity advocacy and feminist hacking is diffuse and could be drawn into greater potency. If these configurations of hacking are not mainly there to produce technological objects, what is their purpose? In this brief conclusion, I trace the proposition of diversity advocacy in open tech outward toward some of its logical conclusions, where as-yet-unclaimed potential becomes more evident. Four interrelated lines of possibility stand out:

First, diversity in tech is a very limited way of claiming equal rights. Technology has gained such dazzling prominence as a site of cultural authority that it seems, to many, that it is a main lever for ensuring political participation writ large. But this book argues that a just society must focus on a multitude of routes to equal social standing, including nontechnological ones. (It seems possibly gratuitous to write these words, but they are worth stating outright.) While it is good for technologists to cultivate their own backyard in terms of equal rights and treatment in sites of technological production, it is a mistake to allow these sites to stand in for the type of democratic participation that a just society would require. Because social power and technological participation can seem so utterly interwoven, it is easy to mistake one for the other, but this is a critical error. Put differently, to frame social inequality as a question of diversity in technological production, and to expect to change wider inequities by adding so-called diverse individuals to technical cultures, is to misunderstand how the distribution of various social identities in a given sector are *outgrowths of differential social power*, not the other way around.

Second, market logics deserve scrutiny here. This book has elevated the voices and social world of those agitating within open-technology communities, but of course their social world shades and fragments into other worlds. Advocates often find their message can proceed with greater ease when they produce rationales for their efforts that dovetail with calls for diversity in industry spaces. One reason for this resonance in rhetoric has to do with open-technology diversity advocates' *adjacency* to interpretive frameworks that are devoted to the profit-oriented pursuit of technological development and growth. When extolling the virtues of diversity in tech, Christen, a thirty-one-year-old German, said in an interview, "I cannot buy a bigger smartphone because it won't fit in my pockets [as a woman]. Apple didn't include period tracking in their health app, [and] face recognition software regularly fails people of color."[3] This is a perfect encapsulation of a market logic being articulated by a volunteer diversity advocate: she touches on race; she touches on gender; she touches on products she can buy; and she steers clear of any controversy that surrounds, for example, face recognition software, such as its relationship to surveillance and algorithmic incursions into citizens' compacts with states. But she also disclosed a more expansive notion of what is at stake for her: "I wish, more diversity [in tech] would mean for everyone, who is not a white heterosexual able-bodied male, to finally feel normal and not like a freak. . . . Even if it meant just this bit of respect and humanity it would change the world." Though she invokes respect for difference and "humanity" at the core of her vision for diversity in tech, product-centered explanations of the value of diversity are always rhetorically within reach and always an easy shorthand—but this shorthand shortchanges diversity advocates whose ultimate pursuit is social justice.

As cultural historian Thomas Streeter has noted, within a given interpretive community informal and commonly shared assumptions are rarely exposed in ways that would invite a full range of comment or questioning: "imponderables that otherwise might be open to an infinite variety of interpretation . . . are given relatively stable, agreed-upon meanings."[4] But the

3. Email,—to author, July 2, 2015. See Benjamin 2016. Of course, though it is constitutively racist, facial recognition software is still scary and ripe for abuse even if its algorithms could be made to be "less racist" (see also Noble 2018).

4. Streeter 1996: 115.

point is not that diversity advocates in voluntaristic open-technology communities necessarily or always fully inhabit the worlds where market logics are so agreed upon as to be unworthy of comment. They do not. Rather, the prominence of market-based rationales for technological development means that market logics are always quite easily affixed to diversity advocacy, and other articulations require more effort. Market logics and racialized capitalism become the basis for emphasizing diversity, rendering other potentials more muted—not completely imponderable, but less salient than they might otherwise be.

Advocates also routinely invoke workplace relations, especially the notion that there is an untapped talent pool and latent diverse workforce waiting in the wings to claim just reward in tech industry employment. Examples of this abound. In a group discussion at one of the unconferences for women in open technology, one person urged people making comments to first give their names, a common enough request in a group setting. But she phrased her request, "Please say your name so that someone can say, 'Bess, that was an awesome comment, I'd like to hire you.' "[5] Though it is possible that workplace networking was occurring at the unconference, this was not its stated purpose; this comment should be taken as an almost ritualistic commitment to workplaces as a main object of intervention. This is somewhat curious given that many advocates acknowledge that their efforts do not produce—and may not even intend to produce—people readied for the workplace.

While goals of better products and workplace diversity are both fine as far as they go, they reflect a limited basis for confronting unequal social power in technological development. In addition, these framings are exactly what we would expect within industry settings, including—or especially—ones that claim to want to *make the world a better place,* the oft-lampooned mandate of many Silicon Valley firms. A stereotypical example is a post on *Tech Crunch* (an online publication featuring tech trends and industry news, founded in Silicon Valley in 2005), entitled "40 Diverse People Who Made Big Moves in 2015." It states: "Diverse people—people of color, LGBTQ men and women, women in general, people with disabilities, older people, younger people, etc.—have contributed a lot to the overall tech ecosystem this year."[6] This is representative discourse in many ways: it highlights and celebrates diversity without specifying what, exactly, diversity is good for; and it frames diversity as inhering to individuals and based in demographic categories like race,

5. Fieldnotes, June 8, 2013, San Francisco, CA.
6. Dickey 2015.

gender, age, and (dis)ability status. Detached from "scary" issues (matters of social justice, structural inequality, or the like, evoked by feminist theorist Sara Ahmed), it is corporate- and institution-friendly.[7]

The diversity advocacy featured in this book is *not* occurring in industry settings; it is prominent in noncommercial and often quite amateur spaces. But its resemblance to industry framings is not without consequences, of course. Unfortunately, market logics have the potential to choke civil rights like so many weeds. Most glaringly, such framings lead to the underdevelopment of goods and services that do not obviously have a market to capture, even if they may serve a public good. And they relegate technological decision-making to elites pursuing profit motives; they are essentially antidemocratic, and the sociotechnical arrangements they usher into being are not easily undone.[8] Thus to let the interpretive frame of market value provide the rationale for diversity advocacy is to severely limit its potential for those who are ostensibly pursuing other values, like justice and equity.

Market framings also leave unexamined some of the power dynamics that advocates' efforts nominally intend to critique. For example, though advocates often claim that a more diverse workforce is partly about members of minoritized or disadvantaged groups claiming some of the benefit that tech industry employment is often held to provide, this ignores the reality of historical struggles between labor and capital. Coding as a new literacy invites comparison to the nineteenth century, when, according to labor scholar Ursula Huws, industry, national economies, and empires became complex enough to demand a workforce that possessed numeracy and literacy.[9] She argues that when these skills became relatively universal, no one in the workforce could exert extra leverage in the market. The point is not that some should have remained illiterate in order to maximize worker power for a few, of course. Instead, there may be parallels to the present in that universal or near-universal attainment of skills like coding will not necessarily enhance worker power unless they are accompanied by regulations and healthy, accountable institutions that would help them gain this power,

7. Ahmed 2012: 66.

8. Philosopher of technology Andrew Feenberg writes, "An undemocratic technical system can offer privileges to its technical servants that might be threatened by a more democratic system. . . . The most important means of assuring more democratic technical representation remains transformation of the technical codes and educational processes through which they are inculcated" (1999: 143).

9. Huws 2014: 40.

such as enforcement of equal pay for equal work, and unions.[10] (Encouraging movement in this regard can be seen in the Tech Workers Coalition, which has been active in organizing both high- and low-status workers in the tech industry;[11] in the Google walkout of late 2018, in which employees walked out to protest discrimination and harassment;[12] and in a nascent movement for worker-owned platform cooperatives in the so-called *sharing economy*.[13]) And yet framings that are especially beneficial to industry—and are more ambiguously beneficial to people oriented to open technology for other purposes—have arguably captured much of voluntaristic diversity advocacy in open technology, which may deflate or stymie the wider emancipatory goals of many advocates.[14]

Third, there are disadvantages to framing diversity in open technology as primarily oriented toward matters of representation. Representation by itself is too indeterminate and not necessarily tied to emancipatory outcomes if it is not anchored to other sorts of politics (as discussed in chapter 6). FLOSS's commitment to articulating liberal ideals and notions of expressive individual sovereignty have made it challenging for its practitioners to fully acknowledge how historical realities of race, class, and gender comprise systems for social sorting and have shaped their collectivities. Diversity advocates have a sharper analysis of these dynamics, readily zeroing in on wider patterns of social difference as they are brought to bear in their communities, and issuing challenges to the liberalism of mainstream FLOSS.

But though abstract rights and freedoms can be extended to new subjects, and difference can be celebrated, as suggested by Herman Gray, diversity advocacy is tripped up on tensions between articulations of diversity that evoke and endorse market logics, and those attuned to deeper and more

10. See Hicks 2018; Salehi 2017; Young 2017.

11. See for example Tarnoff 2017.

12. The walkout organizers encouraged solidarity among not only full-time workers, but contract and temporary workers, who are more likely to be people of color, immigrants, and people from working class backgrounds ("Google Walkout for Real Change," *Medium.com*, November 2, 2018, https://medium.com/@GoogleWalkout/google-employees-and-contractors-participate-in-global-walkout-for-real-change-389c65517843). Discouragingly, mainstream coverage elided this aspect of the protest, highlighting gender discrimination and harassment among higher-status employees like software engineers (Wakabayashi et al. 2018).

13. Scholz and Schneider 2017.

14. Söderberg and Delfanti theorize this as an ongoing dynamic of firms' co-optation of "from below" hacking energies (2015).

heavily politicized concerns.[15] In particular, *representation* can straddle, but not reconcile, diversity advocacy framings that orient around social justice *and* those that seek to expand market potentials (commonly framed as "diverse developers will result in serving/capturing a diverse consumer market," a discourse strongly emanating from corporations). For formerly marginalized or minoritized groups to achieve recognition *as a market* is not a fully unalloyed good, as it may sever production and consumption from other iterations of rights, identity, and visibility.[16] Thus while a discourse of representation is convenient, it is possibly working against some of the emancipatory politics favored by diversity advocates. Like *diversity* itself, representation cannot address fully the "scary" concepts like power and inequity that lie at the core of some advocates' concern.[17] Justice requires not only recognition but redistribution; neither alone is sufficient. It is thus helpful for advocates to articulate and bound their concern.

Representation as a goal may also result in accepting (and reproducing) notions of fixity in terms of social identity. This should raise skepticism, as diversity advocates are, in other contexts, wary of reifying essentialist lines of social difference. Social identity categories are more fruitfully understood as always contingent and constructed, "intrinsically incomplete and open-ended"[18] as opposed to fixed, let alone "natural." Anthropologist Sareeta Amrute describes an insidious dynamic in tech work in the context of migrant Indian coders working in Europe: "Racial difference is prized as a source of new ideas even while it becomes an alibi for differential treatment of temporary workers."[19] This is certainly not a result that diversity advocates would seek, but it is an outgrowth of "recognition and celebration" of difference accompanied by the downplaying of the salience of race and gender *as bases of inequality*, against which Gray warns. There are a variety of reasons to believe that commitments to identity-based representation do not unambiguously amplify political potential.

Lastly, vaunting representation (with an emphasis on difference) as a goal may undercut the cultivation of *coalition and solidarity across difference*. As one unconference attendee had so poignantly asked, "does talking about differences . . . reinforce them?" I do not mean to elide difference, or to suggest that there are not worthwhile tensions and differentials to

15. Fraser 2013: 173.
16. Sender 2005.
17. Ahmed 2012: 66.
18. Wynne 1996: 42; see also Gilroy 2019; Markell 2003; Strathern 2005.
19. Amrute 2018.

be explored in diversity advocacy or other forms of social analysis about equity and justice. Indeed, as Pala, a Belgian designer, said in an interview, "Corporate culture is about taking away tension. [But FLOSS and hacking] projects need to include difference. Tension is different from fear [which we don't want], but there's always misunderstanding, awkwardness. . . . Can we all have the space to speak and differ?"[20] She believed that "tension" could be a productive force for members of projects committed to emancipatory politics. Feminist scholars and others have argued that it is preferable to construct coalition across differences, capturing the generative potential of recognizing difference without being reduced to—or by—that recognition. (This approach also forestalls segmentation into ever-smaller identity-based groups.) Many women* open technologists commenting on gender claimed that men could be allies and buddies. Some women of color said that their solidarity and trust with white women was often quite complicated. These are obvious—almost facile—examples, but they underscore how emancipatory kin relations ought not be reduced to predetermined social identity categories. To regard social identities as "intrinsically incomplete and open-ended" leaves room to forge new categories of belonging and alliance. Pushing past representation as a goal offers the possibility to tap into more political potential in coalition across difference, without seeking to obscure difference.

Fourth, diversity advocacy in open technology faces unique challenges to the extent that it centers on technology as an orienting concept. Technology has a unique stature in our culture. It stands in for things greater than artifacts; it is understood to have profound effects on social order. Conversations about technology are rarely about *artifacts in themselves*. Technology's stature is part of what draws people to it to enact social change. But the social relations and historical patterns that surround it are always freighted and often reflect the priorities and interests of groups with greater social power, such as elites, technocrats, and (within a capitalist system) corporations. According to historian Leo Marx, technology is a "hazardous concept": it fills a semantic and conceptual void in our culture, wherein technology is vested with the power to determine the course of human events all by itself.[21] Thus it is always challenging to engage with this concept in a way that accounts for its presence and power accurately and with appropriate nuance.

20. Interview, Pala, New York, NY, April 26, 2014. Pala's quote also highlights the questions for FLOSS and hacking groups over when it is productive to "fork" versus to keep hacking together, even when differences arise. See Arvin et al. 2013 on crafting alliances that address difference.

21. Marx 2010.

These matters are paramount when considering diversity advocacy in open-technology cultures. All open technologists (not only diversity advocates) are oriented around open technology as a site for expression of agency and a force for social change; in Anika's words, "Open tools are powerful."[22] Indeed, it is productive to understand technologies in general as "thought-objects for the collective enactment and exploration of hopes, desires and political visions."[23] Thus, conversations about who participates—the need to consider diversity—carry baggage, often implicit, about a much bigger progressive project. As stated above, these discussions implicitly refer to social goods, including democratic participation, agency over technology, and often social justice; they are not merely about technological production. This is why it is necessary to carefully tease apart these concepts. *Technology* cannot stand in for implicit or abstract social goods *because of how it is implicated in social relations and unequal power*. Without attention to the hazards of technology, even sowing greater diversity in open-technology cultures may fail to reset the power dynamics advocates aim to address, leaving the power structures surrounding technology essentially intact. In other words, technological intervention needs to be accompanied by a critical analysis of technology that names and seeks to redress imbalances of power.

We urgently need to engage in *explicit* political work and formal political intervention, not a prefigurative politics of techno-utopianism in which social problems are solved by expanding our technical practitioner base. In other words, the notion of "building up young people of color in tech so that we can finally tackle structural inequity," as urged by Dream Corps, has the cart before the horse. This techno-politics fails to acknowledge its own complicity in sorting people into insiders with social power and those set outside. As historian of engineering Amy Slaton writes, "the delineation of STEM learning and work from other cultural projects [is] itself a highly efficient instrument of discrimination."[24] It is not possible to shift the balance of power by bringing new people in without examining the whole arrangement. And hoping for social change without confronting this legacy more directly creates a missed opportunity to intervene into open-technology cultures.

This is also part of why it is perilous to frame diversity advocacy as a project of worker empowerment. As STS scholar Virginia Eubanks has shown, social location sorts people into more or less desirable jobs, and

22. Interview, Anika, July 7, 2015, New York, NY.

23. Streeter 2017.

24. Slaton 2017.

then job categories are used to *selectively* construct the notion of IT work as favorable. Women in particular (but not only women) holding jobs in the bottom portion of an increasingly bifurcated economy do participate in the high-tech economy by supporting those toward the top.[25] In addition, empirically, much IT work is itself low-status and precarious, and training ever more people to be programmers is likely to exacerbate this, not shift it favorably for workers (especially if we take the global political economy into account).[26] Thus, Eubanks recommends shifting focus away from high-tech *products and skills* and argues that we need to analyze and intervene into the interlocking issues that can move us toward a "high-tech equity agenda."[27] While many diversity advocates have excellent intentions and no doubt intend for their efforts to lead to positive outcomes for relatively disempowered workers, cutting corners with the analysis—letting *diversity in tech* stand in for an equity agenda—is bound to lead to suboptimal results, failing to upend the status quo.

This is not to reject technology—it is part of our society (an instantiation of our society), and we should engage it. But this should occur in a mindful way that recognizes limits of technology as a project for empowerment.[28] Of major consequence is the creeping mandate of Silicon Valley, whereby we are all urged to have programming skills, lest we become irrelevant in the brave new future. This threatens to dilute worker power, as noted above. Even more than that, it undermines our ability to solve social problems using the full range of tools at our disposal. In other words, not only should we push back on the notion that all *must* to learn to code, we should also push for geeks and technocrats to *learn more social theory and history*. "Making the world a better place" simply *cannot happen* without rigorous social analysis that includes in its purview technologists and their own social positions. We should immediately ask, *Whose world? Better for whom? Cui bono?*[29]

To the extent that technical cultures and artifacts *are* an object of progressive intervention, it is worthwhile to be reflective about what inequities they can and cannot overcome. It is especially valuable to be conscientious about blind spots in technology-centered interventions and to take into account the structural and historical reasons why technical cultures have been instrumental in maintaining the global and local dominance of elites,

25. Eubanks 2012: 75.
26. Freeman 2000; Amrute 2016; see also Poster and Wilson 2008.
27. Eubanks 2012: 126.
28. Thanks to Chris Csíkszentmihályi for discussion on this.
29. Star 1991.

technocrats, and capitalists. To intervene into these power relations includes reassessing the primacy of tech and questioning how technical engagement and technical communities are themselves bounded. This is different from adding different kinds of bodies to unexamined technical cultures. Rather, it involves a larger reevaluation and appropriation of categories themselves—the boundaries of what is "social" and what is "technical" are flexible categories that have historically constituted resources for claiming "neutrality" while separating and sorting people—in order to build (socially, technically, and analytically) in new directions.

If the goal for grassroots organizers in open-technology communities is less about preparing people for workplaces and more about other inchoate outcomes, it is worthwhile to name those outcomes and work toward them. Most of the people whose efforts I chronicle in this book *are* sincerely invested in some form of "making the world a better place," even if they would give the side-eye to many of the corporate practices of prominent tech companies like Uber or Facebook (and even to Sheryl Sandberg and her "lean in" feminism). But as Clara astutely points out in chapter 6, advocates themselves can wind up muddled when there is encouragement for diversity in their workplaces that stops short of a fuller social justice mandate. When "advocates of course learn the language that will get an initiative funded, or whatever their goal is" (in her words), this is *realpolitik*. But the muddle can potentially dilute their energies. This underscores why wicked (and "scary") social problems cannot be solved by adopting the workplace logics with which they overlap. Though voluntaristic open-technology cultures are in many ways contiguous with workplace cultures, part of their appeal is to allow people spaces to hack and work on projects of authentic affective interest or pleasure to themselves. The impulse to open up these spaces or these feelings to others has plainly never been primarily about the workplace; more is at stake in these collective and individual experiences of joy, problem-solving, and agency.

Voluntaristic technology communities are important sites because they are utopian spaces where people play and tinker not only with technical artifacts but with social reality, imagining social relations through participation in a third space outside work and home, though they are in dialogue and tension with labor markets and domestic economies. (Python in particular is feeding into the market in obvious ways.) However, a hacker solution to the perception of unequal participation the ranks of open technology—a hands-on rough consensus and running code, or, in diversity advocacy, a DIY infrastructure of care—is, on its own, almost certain to come up short.

The challenges of changing social structure or dismantling systemic inequity are too big of an ask for DIY communities. Crucially, this is not actually a shortcoming of voluntaristic communities; building social change requires sustained efforts on multiple fronts, and it is not fair or realistic to ask voluntaristic efforts to bootstrap solutions to inequities. The shortcoming would be to misidentify the matter of *unequal participation in tech* as a *tech problem* with a *tech solution*—or even a tech community problem, with a tech community solution. Such framings obscure the real social relations that lie behind and produce technology itself (and technical communities)—that is, wider formations of social difference.[30] (It is also true that in voluntaristic communities and in workplaces, social norms of politeness mean that bringing to the surface controversial or challenging topics is often fraught with interpersonal peril and tacitly discouraged, as discussed in chapter 6.)

How advocates draw borders around their "matters of care" is important.[31] Bringing borders of care into focus can help clarify just what the discharge of political energies in these sites is actually *meant to do*. If it is to promote justice or an equity agenda, that may well draw attention *away from* tech and toward more elemental threads of the social fabric in a just society. Instead of aiming at diversity in tech, political energies could instead be directed toward desegregating our communities, pursuing fairer distribution of wealth, shoring up public and educational institutions (including libraries), decolonizing institutions, and finding alternatives to shareholder capitalism and attendant rampant ecological destruction. The latter are bigger challenges, needless to say, but they are where a more holistic and less defanged notion of diversity advocacy could very well lead. Attaining them would, incidentally, likely lead to changes in techno-cultures, including who participates in open technology.[32] Equal voluntaristic participation (or equal representation in career paths) are outcomes of a just society, not the prize itself; the former cannot be retrofitted to engineer the latter. To the extent that technical engagement is a part of this project, it might start from a premise of "generative justice,"[33] as opposed to technical participation or diversity in tech. Diversity is necessary, but not sufficient; it represents a shortcut in what should be a deeper conversation about values and justice.

30. As historian of technology David Noble has written, "behind the technology that affects social relations lie the very same social relations" (quoted in Wajcman 2015: 90).

31. Puig de la Bellacasa 2011.

32. Eglash et al. 2004.

33. Eglash 2016.

Diversity advocates can take heart in the fact that the problem is not only how to open up hacking to new sorts of people. Hacking at the periphery has not enjoyed the attention that more mainstream practices like FLOSS have. But if we take a series of situated views from outside this cultural mainstream, it is clear that hacking has never been centered exclusively around white men in the Global North.[34] Furthermore, some of what is required here is simply to shift the frame of what counts as hacking: to redraw boundaries to *place social and historical analysis and infrastructural care work within the purview of hacking*. In combination, these analytical adjustments can illuminate the "others" hacking—who are already here.

To reiterate, the politics of diversity advocacy to date is often indeterminate. But rather than saying it is therefore "not productive," we might ask, what is it productive of? What this book has shown is that diversity in open-technology advocacy exposes a complex dialectic between social relations that reinscribe a dominant order and others that begin to challenge and reconfigure this order. Technologists are involved in cultural mediation as much as technological production.[35] Technology is, if not always already of the dominant culture, always laden with a legacy of division. Technical cultures are world-making: they sort people and present barriers to entry by design. Diversity advocates recognize these dynamics and challenge them, but often incompletely. This book suggests that though advocates might tinker on the margins, such tinkering does not change the fact that technology as an edifice is freighted with a legacy of having been built, in the first place, to shore up the positions of elite and powerful entities. A focus on technology *itself* (or even technical cultures in relative isolation) may be confining to a social justice agenda.[36] It might be possible to build more democratic technology—undoubtedly, it *is* possible—but at the same time, democratic praxis should never be limited to a technological imaginary.

34. E.g., Eglash and Bleecker 2001; Nguyen 2016; Toupin 2016b.

35. Marvin 1988: 7.

36. Or as Judy Wajcman writes, "[P]olitics and not technology per se is the key to . . . equality" (2007: 287).

BIBLIOGRAPHY

Abbate, Janet. "Code Switch: Alternative Visions of Computer Expertise as Empowerment from the 1960s to the 2010s." *Technology and Culture* 59, no. 4 (2018): S134–S159.

Abbate, Janet. *Recoding Gender: Women's Changing Participation in Computing.* Cambridge, MA: MIT Press, 2012.

AdaCamp. "AdaCamp Toolkit."Accessed May 25, 2016. https://adacamp.org/adacamp-toolkit/website-content/.

Ahmed, Sara. *On Being Included.* Durham, NC: Duke University Press, 2012.

Ames, Morgan, Jeffrey Bardzell, Shaowen Bardzell, Silvia Lindtner, David Mellis, and Daniela Rosner. "Making Cultures: Empowerment, Participation, and Democracy—or Not?" *CHI Extended Abstracts on Human Factors in Computing Systems* (2014): 1087–92.

Amrute, Sareeta. *Encoding Race, Encoding Class: Indian IT Workers in Berlin.* Durham, NC: Duke University Press, 2016.

———. "Proprietary Freedoms in an IT Office: How Indian IT Workers Negotiate Code and Cultural Branding." *Social Anthropology* 22, no. 1 (2014): 101–17.

———. "What Would a Techno-Ethics Look Like?" Platypus: The CASTAC (blog). Last modified 16, 2018. http://blog.castac.org/2018/01/techno-ethics/.

Anarchaserver. Servidor Feminista/Feminist Server. September 26, 2014. http://anarchaserver.org/mediawiki/index.php/Servidor_feminista/Feminist_Server.

Aouragh, Miriyam, and Paula Chakravartty. "Infrastructures of Empire: Towards a Critical Geopolitics of Media and Information Studies." *Media, Culture & Society* 38, no. 4 (2016): 559–75.

Arvin, Maile, Eve Tuck, and Angie Morrill. "Decolonizing Feminism: Challenging Connections between Settler Colonialism and Heteropatriarchy." *Feminist Formations* 25, no. 1 (2013): 8–34.

Asaro, Peter. "On Banning Autonomous Weapon Systems: Human Rights, Automation, and the Dehumanization of Lethal Decision-Making." *International Review of the Red Cross* 94, no. 886 (2012): 687–709.

Aspray, William. *Women and Underrepresented Minorities in Computing: A Historical and Social Study.* Basel, Switzerland: Springer, 2016.

Bailey, Cameron. "Virtual Skin: Articulating Race in Cyberspace." In *Immersed in Technology: Art in Virtual Environments*, edited by Mary Ann Moser. Cambridge, MA: MIT Press, 1996.

Baker, Katie. "Meet the Brogrammers." *Jezebel* March 3, 2012. https://jezebel.com/5890224/meet-the-brogrammers.

Banet-Weiser, Sarah. *Authentic™: The Politics of Ambivalence in a Brand Culture.* New York: New York University Press, 2012.

———. "'Confidence You Can Carry!': Girls in Crisis and the Market for Girls' Empowerment Organizations." *Continuum* 29, no. 2 (2015): 182–93.

Banet-Weiser, Sarah, and Kate M. Miltner. "#MasculinitySoFragile: Culture, Structure, and Networked Misogyny." *Feminist Media Studies* 16, no. 1 (2016): 171–74.

Banks, David. "The Conservative Hacker." Cyborgology. Last modified August 26, 2015. https://thesocietypages.org/cyborgology/2015/08/26/the-conservative-hacker/.

Beard, Mary. "The Public Voice of Women." *London Review of Books* 36, no. 6 (2014): 11–14.

Beaulieu, Anne, and Adolfo Estalella. "Rethinking Research Ethics for Mediated Settings." *Information, Communication & Society* 15 (2011): 1–20.

Bell, David. "Geek Myths: Technologies, Masculinities, Globalizations." In *Rethinking Transnational Men*, edited by Jeff Hearn, Marina Blagojević, Katherine Harrison. New York: Routledge, 2013.

Beltrán, Héctor, Jr. "Hacking Imaginaries: Codeworlds and Code Work across the U.S./Mexico Borderlands." PhD dissertation, University of California, Berkeley, 2018.

Benjamin, Ruha. "Innovating Inequity: If Race Is a Technology, Postracialism Is the Genius Bar." *Ethnic and Racial Studies* 39, no. 13 (2016): 2227–34.

Benkler, Yochai. *The Wealth of Networks: How Social Production Transforms Markets and Freedom.* New Haven, CT: Yale University Press, 2006.

Benner, Chris. "Learning Communities in a Learning Region: The Soft Infrastructure of Cross-Firm Learning Networks in Silicon Valley." *Environment and Planning A: Economy and Space* 35, no. 10 (2003): 1809–30.

Bix, Amy Sue. *Girls Coming to Tech! A History of American Engineering Education for Women.* Cambridge, MA: MIT Press, 2014.

Boczkowski, Pablo, and Leah A. Lievrouw. "Bridging STS and Communication Studies: Scholarship on Media and Information Technologies." In *New Handbook of Science, Technology and Society,* edited by E. Hackett, O. Amsterdamska, M. Lynch, and J. Wajcman, 949–78. Cambridge, MA: MIT Press, 2008.

Borsook, Paulina. *Cyberselfish: A Critical Romp through the Terribly Libertarian Culture of High Tech.* New York: Perseus, 2000.

boyd, danah. "Why America is Self-Segregating." *Points: Data & Society* (blog). Last modified January 5,2017. https://points.datasociety.net/why-america-is-self-segregating-d881a39273ab#.xg9n4fgul.

Bratich, Jack. "The Digital Touch: Craft-Work as Immaterial Labour and Ontological Accumulation." *Ephemera: Theory & Politics in Organization* 10, no. 3–4 (2010): 303–18.

Bruckman, Amy. "Studying the Amateur Artist: A Perspective on Disguising Data Collected in Human Subjects Research on the Internet." *Ethics and Information Technology* 4, no. 3 (2002): 217–31.

Buechley, Leah, and Benjamin Mako Hill. "LilyPad in the Wild: How Hardware's Long Tail Is Supporting New Engineering and Design Communities." In *Proceedings of the 8th ACM Conference on Designing Interactive Systems*, 199–207. ACM, Aarhus, Denmark, August 16–20, 2010.

Butler, Judith. *Bodies That Matter.* New York: Routledge, 1993.

———. *Gender Trouble.* New York: Routledge, 1990.

Byfield, Bruce. "The Ada Initiative Leaves a Mixed Record behind It." *Off the Beat, Bruce Byfield's Blog.* Last modified August 14, 2015. http://www.linux-magazine.com/Online/Blogs/Off-the-Beat-Bruce-Byfield-s-Blog/The-Ada-Initiative-leaves-a-mixed-record-behind-it.

———. "Debian Women: Geek Feminists in Action." Linux.com. Last modified December 8, 2004. https://www.linux.com/news/debian-women-geek-feminists-action.

———. "It's Time for the Next Step with Anti-Harassment Policies." *Off the Beat, Bruce Byfield's Blog.* Last modified March 22, 2013. http://www.linux-magazine.com/Online/Blogs/Off-the-Beat-Bruce-Byfield-s-Blog/It-s-time-for-the-next-step-with-anti-harassment-policies.

Callahan, Brian Robert, Charles Hathaway, and Mukkai Krishnamoorthy. "Quantitative Metrics for Generative Justice: Graphing the Value of Diversity." *Teknokultura* 13, no. 2 (2016): 567–86.

Cameron, Deborah. "Styling the Worker: Gender and the Commodification of Language in the Globalized Service Economy." *Journal of Sociolinguistics* 4, no. 3 (2000): 323–47.

Cárdenas, Micha, Zach Blas, and Wolfgang Schirmacher. *The Trans Real: Political Aesthetics of Crossing Realities.* New York: Atropos Press, 2012.

Carnegie Mellon University. "CMU's Proportion of Undergraduate Women in Computer Science and Engineering Soars above National Averages." September 12, 2016, https://www.cmu.edu/news/stories/archives/2016/september/undergrad-women-engineering-computer-science.html.

Carrigan, Coleen. "Cracking the Code: Navigating and Subverting Dominant Class Rule in Computer Science and Engineering." PhD dissertation, University of Washington, 2013.

Carse, Ashley. "Keyword: Infrastructure—How a Humble French Engineering Term Shaped the Modern World." In *Infrastructures and Social Complexity: A Routledge Companion*, edited by Penny Harvey, Casper Bruun Jensen, and Atsuro Morita. London: Routledge, 2016.

Ceruzzi, Paul. *A History of Modern Computing.* Cambridge, MA: MIT Press, 2003.

Chachra, Debbie. "Why I Am Not a Maker." *Atlantic*, January 23, 2015. http://www.theatlantic.com/technology/archive/2015/01/why-i-am-not-a-maker/384767.

Chakravartty, Paula, and Mara Mills. "Virtual Roundtable on 'Decolonial Computing.'" *Catalyst: Feminism, Theory, Technoscience* 4, no. 2 (2018). https://doi.org/10.28968/cftt.v4i2.29588.

Chan, Anita. "Decolonial Computing and Networking Beyond Digital Universalism." *Catalyst: Feminism, Theory, Technoscience* 4, no. 2 (2018). https://doi.org/10.28968/cftt.v4i2.29844.

———. *Networking Peripheries: Technological Futures and the Myth of Digital Universalism.* Cambridge, MA: MIT Press, 2013.

Chappell, Ben. "'Take a Little Trip with Me:' Lowriding and the Poetics of Scale." In *Technicolor: Race, Technology and Everyday Life*, edited by Thuy Linh N. Tu and Alondra Nelson. New York: New York University Press, 2001.

Chappell, Bill. "Texas High School Student Shows Off Homemade Clock, Gets Handcuffed." *National Public Radio*, September 15, 2015. http://www.npr.org/sections/thetwo-way/2015/09/16/440820557/high-school-student-shows-off-homemade-clock-gets-handcuffed.

Chen, Adrian. "The Truth about Anonymous's Activism." *Nation*, November 11, 2014.

Chen, Katherine. *Enabling Creative Chaos: The Organization behind the Burning Man Event.* Chicago: University of Chicago Press, 2009.

Chess, Shira and Adrienne Shaw. "A Conspiracy of Fishes, or, How We Learned to Stop Worrying About #GamerGate and Embrace Hegemonic Masculinity." *Journal of Broadcasting & Electronic Media* 59, no. 1 (2015): 208–220.

Child Soldiers International. "International Laws and Child Rights." Accessed August 19, 2017. https://www.child-soldiers.org/international-laws-and-child-rights.

Christian, Aymar J. *Open TV: Innovation beyond Hollywood and the Rise of Web Television.* New York: New York University Press, 2018.

Chua, Mel. "About." *Mel Chua: Hacker. Writer. Researcher. Teacher. Human jumper cable.* Accessed December 31, 2016. http://blog.melchua.com/about/.

———. "On the Diversity-Readiness of STEM Environments: 'It's Almost As If I Could Only Enter the Makerspace As a Janitor.'" *Mel Chua: Hacker. Writer. Researcher. Teacher. Human jumper cable* (blog). Last modified April 28, 2015. http://blog.melchua.com/2015/04/28/on-the-diversity-readiness-of-stem-environments-its-almost-as-if-i-could-only-enter-the-makerspace-as-a-janitor/.

Chun, Wendy Hui Kyong. "Race and/as Technology, Or How to Do Things to Race." In *Race after the Internet*, edited by Lisa Nakamura and Peter Chow-White. New York: Routledge, 2011.

Coleman, E. Gabriella. *Coding Freedom*. Princeton, NJ: Princeton University Press, 2012a.

———. "From Internet Farming to Weapons of the Geek." *Current Anthropology* 58, no. S15 (2017): S91-S102.

———. "The Hacker Conference: A Ritual Condensation and Celebration of a Lifeworld." *Anthropological Quarterly* 83, no. 1 (2010): 47–72.

———. *Hacker, Hoaxer, Whistleblower, Spy: The Many Faces of Anonymous*. London: Verso: 2015.

———. "Phreakers, Hackers, Trolls: The Politics of Transgression and Spectacle." In *The Social Media Reader*, edited by Michael Mandiberg. New York: New York University Press, 2012b.

Collins, Patricia Hill. *Black Feminist Thought: Knowledge, Consciousness, and the Politics of Empowerment*. New York: Routledge, 2000.

Combahee River Collective. (1977) "A Black Feminist Statement." *Women's Studies Quarterly* 42, no. 3–4 (2014): 271–80.

Connell, R.W. *Masculinities*. Sydney: Allen & Unwin, 2005.

Corneliussen, Hilde. *Gender-Technology Relations: Exploring Stability and Change*. New York: Palgrave Macmillan, 2013.

Costanza-Chock, Sasha. *Design Justice*. Cambridge, MA: MIT Press, 2019.

Couture, Stéphane, and Sophie Toupin. "What Does the Concept of 'Sovereignty' Mean in Digital, Network and Technological Sovereignty?" Global Internet Governance Academic Network Conference, Geneva, Switzerland, 2017.

Crenshaw, Kimberlé. "Mapping the Margins: Intersectionality, Identity Politics, and Violence against Women of Color." *Stanford Law Review* 43, no. 6 (1991): 1241–99.

Cryptodance. n.d. Accessed February 17, 2018. http://www.ooooo.be/cryptodance/index.html.

Dafermos, George, and Johan Söderberg. "The Hacker Movement as a Continuation of Labor Struggle." *Capital & Class* 33, no. 1 (2009): 53–73.

Dame, Avery. "Mapping the Territory: Archiving the Trans Website in an Age of Search." *Transgender Studies Quarterly* 3, no. 3–4 (2016): 628–36.

Daniels, Jessie. "Race and Racism in Internet Studies: A Review and Critique." *New Media & Society* 15, no. 5 (2013): 695–719.

———. "The Trouble with White Feminism: Whiteness, Digital Feminism and the Intersectional Internet." In *The Intersectional Internet: Race, Sex, Class, and Culture Online*, edited by Safiya Noble and Brendesha M. Tynes. New York: Peter Lang International Academic Publishers, 2016.

Davies, Sarah. "Interviews with Generation Y: Willow Brugh." August 19, 2009. http://sarahdavies.cc/2009/08/19/interviews-with-generation-y-willow-brugh/.

Davies, Sarah R. "Characterizing Hacking: Mundane Engagement in US Hacker and Makerspaces." *Science, Technology, & Human Values* (2017a). https://doi.org/10.1177%2F0162243917703464.

———. *Hackerspaces: Making the Maker Movement*. Malden, MA: Polity Press, 2017b.

———. "On Being Inconsequential: Making, Craft and Liquid Leisure." Manuscript draft in the author's possession, 2017c.

Davis, Angela Y. *Women, Race and Class*. New York: Random House, 1981.

Delfanti, Alessandro, and Söderberg, Johan. "Repurposing the Hacker: Three Cycles of Recuperation in the Evolution of Hacking and Capitalism." SSRN. June 23, 2015. https://ssrn.com/abstract=2622106.

Delgado, Richard and Jean Stefancic. *Critical Race Theory*. New York: New York University Press, 2001.

Dickey, Megan Rose. "40 Diverse People Who Made Big Moves in 2015." *Tech Crunch* December 23, 2015. https://techcrunch.com/gallery/40-diverse-people-in-tech-2015/.

Douglas, Susan. *Inventing American Broadcasting*. Baltimore: Johns Hopkins University Press, 1987.

Downey, Greg. *Telegraph Messenger Boys.* New York: Routledge, 2002.

———. "Virtual Webs, Physical Technologies, and Hidden Workers: The Spaces of Labor in Information Internetworks." *Technology and Culture* 42 (2001): 209–35.

Duncombe, Stephen. *Dream: Re-Imagining Progressive Politics in an Age of Fantasy.* New York: New Press, 2007.

Dunbar-Hester, Christina. "Feminists, Geeks, and Geek Feminists: Understanding Gender and Power in Technological Activism." In *Media Activism,* edited by Victor Pickard and Guobin Yang, Routledge, 2018.

———. "Frailties at the Borders: Stalled Activist Media Projects in East Africa." *International Journal of Communication* 10 (2016b): 2157–78.

———. " 'Freedom from Jobs' or Learning to Love to Labor? Diversity Advocacy and Working Imaginaries in Open Technology Projects." *Teknokultura* 13, no. 2 (2016c): 541–66.

———. "Geek." In *Digital Keywords,* edited by Ben Peters. Princeton, NJ: Princeton University Press, 2016a.

———. "If 'Diversity' Is the Answer, What Is the Question? Understanding Diversity Advocacy in Open Technology Projects." In *digitalSTS Handbook,* edited by Janet Vertesi and David Ribes. Princeton, NJ: Princeton University Press, 2019.

———. *Low Power to the People: Pirates, Protest, and Politics in FM Radio Activism.* Cambridge, MA: MIT Press, 2014.

———. "What's Local? Localism as a Discursive Boundary Object in Low-Power Radio Policymaking." *Communication, Culture & Critique* 6, no. 4 (2013): 502–24.

———, and Bryce Renninger. "Trans Technology: Circuits of Culture, Self, Belonging." Institute for Women and Art, Rutgers University, Spring 2013.

———, and Gabriella Coleman. "Engendering Change? Gender Advocacy in Open Source." *Culture Digitally: Examining Contemporary Cultural Production* 26 (2012). http://culturedigitally.org/2012/06/engendering-change-gender-advocacy-in-open-source.

DuPont, Quinn and Alana Cattapan. "Engendering Alice and Bob." Backchannels, *Society for Social Studies of Science,* October 9, 2017. https://www.4sonline.org/blog/post/engendering_alice_and_bob.

Edwards, Paul N. "The Army and the Microworld: Computers and the Politics of Gender Identity." *Signs: Journal of Women in Culture and Society* 16, no. 1 (1990): 102–27.

———. *The Closed World: Computers and the Politics of Discourse in Cold War America.* Cambridge, MA: MIT Press, 1996.

———. "Infrastructure and Modernity: Force, Time, and Social Organization in the History of Sociotechnical Systems." In *Modernity and Technology,* edited by Thomas J. Misa, Philip Brey, and Andrew Feenberg. Cambridge, MA: MIT Press, 2003.

Eglash, Ron. *African Fractals.* New Brunswick, NJ: Rutgers University Press, 1999.

———. "An Introduction to Generative Justice." *Teknokultura* 13, no. 2 (2016): 369–404.

———. "Race, Sex, and Nerds: From Black Geeks to Asian American Hipsters." *Social Text* 71, no. 2 (2002): 49–64.

———, and David A. Banks. "Recursive Depth in Generative Spaces: Democratization in Three Dimensions of Technosocial Self-Organization." *Information Society* 30, no. 2 (2014): 106–15.

———, and Julian Bleecker. "The Race for Cyberspace: Information Technology in the Black Diaspora." *Science as Culture* 10 (2001): 353–74.

———, Jennifer Croissant, Giovanna Di Chiro, and Rayvon Fouché, eds. *Appropriating Technology: Vernacular Science and Social Power.* Minneapolis: University of Minnesota Press, 2004.

Ellcessor, Elizabeth. *Restricted Access: Media, Disability, and the Politics of Participation.* New York: New York University Press, 2016.

Ensmenger, Nathan. " 'Beards, Sandals, and Other Signs of Rugged Individualism': Masculine Culture within the Computing Professions." *Osiris* 30, no. 1 (2015): 38–65.

———. *The Computer Boys Take Over*. Cambridge, MA: MIT Press, 2010a.

———. "Making Programming Masculine." In *Gender Codes,* edited by Thomas Misa. Hoboken, NJ: Wiley & Sons, Inc/ IEEE Computer Society, 2010b.

Eubanks, Virginia. *Digital Dead End.* Cambridge, MA: MIT Press, 2012.

———. "Double-Bound: Putting the Power Back into Participatory Research." *Frontiers: A Journal of Women's Studies* 30, no. 1 (2009): 107–37.

Evans, Claire. "An Oral History of the First Cyberfeminists." *Motherboard*, December 11, 2014. https://motherboard.vice.com/en_us/article/z4mqa8/an-oral-history-of-the-first-cyberfeminists-vns-matrix.

———. " 'We Are the Future Cunt': Cyberfeminism in the 90s." *Motherboard* November 20, 2014. https://motherboard.vice.com/en_us/article/4x37gb/we-are-the-future-cunt-cyberfeminism-in-the-90s.

Everett, Anna, and Amber Wallace. *Afro-Geeks: Beyond the Digital Divide.* Santa Barbara, CA: Center for Black Studies Research, 2007.

Faulkner, Wendy. " 'Nuts and Bolts and People': Gender-Troubled Engineering Identities." *Social Studies of Science* 37, no. 3 (2007): 331–56.

———. "The Power and the Pleasure? A Research Agenda for 'Making Gender Stick' to Engineers." *Science, Technology, & Human Values* 25, no. 1 (2000): 87–119.

———. "Strategies of Inclusion: Gender and the Information Society" Final Report (Public Version). University of Edinburgh, 2004. http://www.sigis-ist.org.

Federici, Silvia. *Caliban and the Witch.* Brooklyn, NY: Autonomedia, 2004.

Feenberg, Andrew. *Questioning Technology*. New York: Routledge, 1999.

Fenton, Tom. "Explaining Virtualization to Your Mom in 5 Easy Steps." Last modified October 8, 2014. https://virtualizationreview.com/articles/2014/10/08/candy-beer-and-virtualization.aspx.

Forlano, Laura. "Hacking the Feminist Disabled Body." *Journal of Peer Production* 8 (2016). http://peerproduction.net/issues/issue-8-feminism-and-unhacking-2/peer-reviewed-papers/issue-8-feminism-and-unhackingpeer-reviewed-papers-2hacking-the-feminist-disabled-body/.

Foucault, Michel. *Discipline and Punish.* New York: Vintage Books, 1995.

Fouché, Rayvon. *Black Inventors in the Age of Segregation.* Baltimore, MD: Johns Hopkins University Press, 2003.

———. "Say It Loud, I'm Black and I'm Proud: African Americans, American Artefactual Culture, and Black Vernacular Technological Creativity." *American Quarterly* 58, no. 3 (2006): 639–61.

Fowler, Martin. "DiversityMediocrityIllusion." January 13, 2015. http://martinfowler.com/bliki/DiversityMediocrityIllusion.html.

Fox, Sarah, Rachel Rose Ulgado, and Daniela Rosner. "Hacking Culture, Not Devices: Access and Recognition in Feminist Hackerspaces." In *Proceedings of the 18th ACM Conference on Computer Supported Cooperative Work & Social Computing,* 56–68. ACM, Vancouver, BC, 2015.

Fraser, Nancy. *Fortunes of Feminism: From State-Managed Capitalism to Neoliberal Crisis.* London: Verso, 2013.

———. "Rethinking the Public Sphere: A Contribution to the Critique of Actually Existing Democracy." *Social Text* 25/26 (1990): 56–80.

Free Software Foundation. "Happy Ada Lovelace Day!" Last modified October 16, 2012. https://www.fsf.org/blogs/community/happy-ada-lovelace-day.

———. "What Is Free Software?" Accessed February 17, 2015. http://www.fsf.org/about/what-is-free-software.

Freeman, Carla. *High Tech and High Heels in the Global Economy: Women, Work, and Pink-Collar Identities in the Caribbean.* Durham, NC: Duke University Press, 2000.

Freeman, Jo. "The Tyranny of Structurelessness." *Berkeley Journal of Sociology* 17 (1972–73): 151–64.

Gajjala, Radhika. "Third-World Critiques of Cyberfeminism." *Development in Practice* 9, no. 5, (1999): 616–19.

———. "Woman and Other Women: Implicit Binaries in Cyberfeminisms." *Communication and Critical/Cultural Studies* 11, no. 3 (2014): 288–92.

Geekfeminism.org. "Language for Trans-Inclusive Events?" March 19, 2014. https://geekfeminism.org/2014/03/19/language-for-trans-inclusive-events/.

Geekfeminism.org. "That Time I Wasn't Harassed at a Conference." August 15, 2013. http://geekfeminism.org/2013/08/15/that-time-i-wasnt-harassed-at-a-conference/.

Geek Feminism Wiki. "About." Accessed December 19, 2009. http://geekfeminism.wikia.com/wiki/Geek_Feminism_Wiki.

Geertz, Clifford. "Deep Hanging Out." *New York Review of Books* 45, no. 16 (1998): 69–72.

Geiger, R. Stuart, and David Ribes. "The Work of Sustaining Order in Wikipedia: The Banning of a Vandal." *Proceedings of the 2010 ACM Conference on Computer Supported Cooperative Work*, Savannah, Georgia, USA, February 6–10, 2010, 117–26.

Gelber, Steven M. "Do-It-Yourself: Constructing, Repairing and Maintaining Domestic Masculinity." *American Quarterly* 49, no. 1 (1997): 66–112.

Genderchangers.org. "International Women's Day 2007." Accessed February 15, 2015. http://genderchangers.org/march.html.

Ghosh, Rishab A., and Ruediger Glott. "Free/Libre/Open Source Software: Policy Support." *FLOSSPOLS: An Economic Basis for Open Standards* 3 (2005). Accessed September 10, 2009. http://www.flosspols.org/deliverables/FLOSSPOLS-D04-openstandards-v6.pdf.

———, Bernhard Krieger, and Gregorio Robles. "Free/Libre and Open Source Software: Survey and Study." Part IV: Survey of Developers. Maastricht: International Institute of Infonomics/Merit (2002).

Gibson, Camille, Shannon Davenport, Tina Fowler, Colette B. Harris, Melanie Prudhomme, Serita Whiting, and Sherri Simmons-Horton. "Understanding the 2017 'Me Too' Movement's Timing." *Humanity & Society* (March 2019). https://doi.org/10.1177%2F0160597619832047.

Gilroy, Paul. "Never Again: Refusing Race and Salvaging the Human." 2019 Holberg Lecture, May 31, 2019. https://www.holbergprisen.no/en/news/holberg-prize/2019-holberg-lecture-laureate-paul-gilroy.

Github.com. "GitHub Puts Open Code of Conduct on Pause." *Hacker News*. Last modified August 8, 2015. https://news.ycombinator.com/item?id=10027332.

———. "What Is NCoC." Accessed July 15, 2016. https://github.com/domgetter/NCoC.

GNU Operating System. "What Is Free Software?" Accessed December 31, 2016. https://www.gnu.org/philosophy/free-sw.en.html.

Golumbia, David. "Cyberlibertarianism: The Extremist Foundations of 'Digital Freedom,'." Talk delivered at Clemson University, September 5, 2013a.

———. "Cyberlibertarians' Digital Deletion of the Left." *Jacobin Magazine* 4 (2013b). https://www.jacobinmag.com/2013/12/cyberlibertarians-digital-deletion-of-the-left/.

Gray, Herman. "Subject(ed) to Recognition." *American Quarterly* 65, no. 4 (2013): 771–98.

Gray, Kishonna. "Solidarity Is for White Women in Gaming." In *Diversifying Barbie and Mortal Kombat: Intersectional Perspectives and Inclusive Design in Gaming*, edited by Yasmin Kafai, Gabriela Richard, and Brandesha Tynes. Pittsburgh, PA: Carnegie Mellon/ETC Press, 2016.

Green, Venus. "Race and Technology: African American Women in the Bell System, 1945–1980." *Technology and Culture* 36, no. 2, Supplement: Snapshots of a Discipline: Selected Proceedings from the Conference on Critical Problems and Research Frontiers in the History of Technology, Madison, Wisconsin (1995): S101-S144.

Gregg, Melissa. "Hack for Good: Speculative Labour, App Development and the Burden of Austerity." *Fibreculture* 25 (2015). http://dx.doi.org/10.15307/fcj.25.186.2015.

———, and Carl DiSalvo. "The Trouble with White Hats." *New Inquiry*, November 21, 2013. http://thenewinquiry.com/essays/the-trouble-with-white-hats/.

Grenzfurthner, Johannes, and Frank Apunkt Schneider. "Hacking the Spaces." 2009 http://www.monochrom.at/hacking-the-spaces/.

Gusterson, Hugh. *Nuclear Rites: A Weapons Laboratory at the End of the Cold War*. Berkeley: University of California Press, 1996.

———. "Studying Up Revisited." *PoLAR: Political and Legal Anthropology Review* 20, no. 1 (1997): 114–19.

Habermas, Jürgen. *The Structural Transformation of the Public Sphere: An Inquiry into a Category of Bourgeois Society*. Translated by Thomas Burger, with the assistance of Frederick Lawrence. Cambridge, MA: MIT Press, 1991.

Hacker, Sally. *"Doing It the Hard Way": Investigations of Gender and Technology*. Boston: Unwin Hyman, 1990.

Halberstam, Jack. *Female Masculinity*. Durham, NC: Duke University Press, 1998.

Hampton, Rachelle. "Which People?" *Slate*, February 13, 2019. https://slate.com/human-interest/2019/02/people-of-color-phrase-history-racism.html.

Haraway, Donna. "The Cyborg Manifesto." In *Simians, Cyborgs, and Women: The Reinvention of Nature*. New York: Routledge, 1991a.

———. "Situated Knowledges." In *Simians, Cyborgs, and Women: The Reinvention of Nature*. New York: Routledge, 1991b.

Harding, Sandra. *Is Science Multicultural? Postcolonialisms, Feminisms, and Epistemologies*. Bloomington: Indiana University Press, 1998.

———. "Just Add Women and Stir?" In *Missing Links*, edited by Gender Working Group, United Nations Commission on Science and Technology for Development. Ottawa, Canada: International Development Research Centre, 1995.

———. *Objectivity and Diversity: Another Logic of Scientific Research*. Chicago: University of Chicago Press, 2016.

Haring, Kristen. *Ham Radio's Technical Culture*. Cambridge, MA: MIT Press, 2006.

Heilman, Madeline, and Tyler Okimoto. "Why Are Women Penalized for Success at Male Tasks? The Implied Communality Deficit." *Journal of Applied Psychology* 92, no. 1 (2007): 81–92.

Henry, Liz. "The Rise of Feminist Hackerspaces and How to Make Your Own." *Model View Culture*, 3 (2013). https://modelviewculture.com/pieces/the-rise-of-feminist-hackerspaces-and-how-to-make-your-own.

Hess, David J. "Technology-and Product-Oriented Movements: Approximating Social Movement Studies and Science and Technology Studies." *Science, Technology, & Human Values* 30, no. 4 (2005): 515–35.

Hicks, Marie. "The Long History behind the Google Walkout." *Verge*, November 9, 2018. https://www.theverge.com/2018/11/9/18078664/google-walkout-history-tech-strikes-labor-organizing.

———. *Programmed Inequality*. Cambridge, MA: MIT Press, 2017.

Hoffman, Anna Lauren. "Data, Technology and Gender: Thinking about (and from) Trans Lives." In *Spaces for the Future: A Companion to Philosophy of Technology*, edited by Joseph Pitt and Ashley Shew, 3–13. New York: Routledge, 2017.

Honeywell, Leigh. "No More Rock Stars: How to Stop Abuse in Tech Communities." Hypatia.ca (blog). Last modified June 21, 2016. https://hypatia.ca/2016/06/21/no-more-rock-stars/.

hooks, bell. *All about Love*. New York: Harper Collins, 2001.

———. "Dig Deep: Beyond *Lean In*." *Feminist Wire*, October 28, 2013. http://thefeministwire.com/2013/10/17973/.

———. *Talking Back: Thinking Feminist, Thinking Black*. Cambridge, MA: South End Press, 1989.

Hughes, Thomas. "The Evolution of Large Technological Systems." In *The Social Construction of Technological Systems: New Directions in the Sociology and History of Technology*, edited by Wiebe Bijker, Thomas Hughes, and Trevor Pinch. Cambridge, MA: MIT Press, 1987.

Humphries, Matthew. "Lockheed Martin Wants a Fingertip-Sized UAV in Every Soldier's Backpack." Geek.com, August 19, 2011. https://www.geek.com/geek-pick/lockheed-martin-wants-a-fingertip-sized-uav-in-every-soldiers-backpack-1414453/.

Huws, Ursula. *Labor in the Global Digital Economy*. New York: Monthly Review Press, 2014.

———. "The Making of a Cybertariat? Virtual Work in a Real World." *Socialist Register* 37, no. 1 (2001): 1–23.

Irani, Lilly. "Justice for Data Janitors." *Public Books*, January 1, 2015. https://www.publicbooks.org/justice-for-data-janitors/.

———, Paul Dourish, and Melissa Mazmanian. "Shopping for Sharpies in Seattle: Mundane Infrastructures of Transnational Design." In *Proceedings of ICIC*. Copenhagen, Denmark, 2010.

———, and Kavita Philip. "Negotiating Engines of Difference." *Catalyst: Feminism, Theory, Technoscience* 4, no. 2 (2018). https://doi.org/10.28968/cftt.v4i2.29841.

———, Janet Vertesi, Paul Dourish, Kavita Philip, and Rebecca E. Grinter. "Postcolonial Computing: A Lens on Design and Development." *Proceedings of the SIGCHI Conference on Human Factors in Computing Systems*. Atlanta, Georgia, USA, April 10–15, 2010: 1311–20.

Jackson, Steve. "Rethinking Repair." In *Media Technologies*, edited by Tarleton Gillespie, Pablo Boczkowski, and Kirsten Foot. Cambridge, MA: MIT Press, 2014.

Jain, S. L. "The Prosthetic Imagination: Enabling and Disabling the Prosthesis Trope." *Science, Technology, & Human Values* 24, no. 1 (1999): 31–54.

Jones, Michael Owen. "Why Folklore and Organization (s)?" *Western Folklore* 50, no. 1 (1991): 29–40.

Jordan, Tim. "A Genealogy of Hacking." *Convergence* 23, no. 5 (2016): 528–44. https://doi.org/10.1177/1354856516640710.

———, and Paul Taylor. "A Sociology of Hackers." *Sociological Review* 46, no. 4 (1998): 757–80.

Joseph, Ralina. *Postracial Resistance: Black Women, Media, and the Uses of Strategic Ambiguity*. New York: New York University Press, 2018.

Juris, Jeffrey. *Networking Futures: The Movements against Corporate Futures*. Cambridge, MA: MIT Press, 2008.

Kafai, Yasmin B., Carrie Heeter, Jill Denner, and Jennifer Y. Sun. *Beyond Barbie® and Mortal Kombat: New Perspectives on Gender and Gaming*. Cambridge, MA: MIT Press, 2008.

Karel, Rob. "The Internet of Things, So Mom Can Understand." *Informatica Blog*. May 8, 2015. https://blogs.informatica.com/2015/05/08/the-internet-of-things-so-mom-can-understand/.

Karanovic, Jelena. "Activist Intimacies: Gender and Free Software in France." Presented to American Anthropological Association Annual Meetings, Philadelphia, PA. December 3, 2009.

Kearney, Mary Celeste. "Pink Technology: Mediamaking Gear for Girls." *Camera Obscura: Feminism, Culture, and Media Studies* 25, no. 2 (74) (2010): 1–39.

Kelty, Christopher. "The Fog of Freedom." In *Media Technologies*, edited by Tarleton Gillespie, Pablo Boczkowski, and Kirsten Foot. Cambridge, MA: MIT Press, 2014.

———. "Geeks, Social Imaginaries, and Recursive Publics." *Cultural Anthropology* 20, no. 2 (2005): 185–214.

———. "There Is No Free Software." *Journal of Peer Production* 3 (2013). http://peerproduction.net/issues/issue-3-free-software-epistemics/debate/there-is-no-free-software/.

———. *Two Bits: The Cultural Significance of Free Software*. Durham, NC: Duke University Press, 2008.

Kendall, Lori. *Hanging out in the Virtual Pub*. Berkeley: University of California Press, 2002.

Khatchadourian, Raffi. "Julian Assange, a Man without a Country." *New Yorker*, August 21, 2017. http://www.newyorker.com/magazine/2017/08/21/julian-assange-a-man-without-a-country.

Kim, Hyomin, Youngju Cho, Sungeun Kim, and Hye-Suk Kim. "Women and Men in Computer Science: Geeky Proclivities, College Rank, and Gender in Korea." *East Asian Science, Technology and Society* 12, no. 1 (2018): 33–56.

Klein, Naomi. *The Shock Doctrine: The Rise of Disaster Capitalism*. New York: Picador, 2007.

Klinenberg, Eric. *Palaces for the People: How Social Infrastructure Can Help Fight Inequality, Polarization, and the Decline of Public Life*. New York: Crown, 2018.

Knorr-Cetina, Karin. *Epistemic Cultures: How the Sciences Make Knowledge*. Cambridge, MA: Harvard University Press, 1999.

Kreiss, Daniel, Megan Finn, and Fred Turner. "The Limits of Peer Production: Some Reminders from Max Weber for the Network Society." *New Media & Society* 13, no. 2 (2011): 243–59.

Kukorowski, Drew. "The Price to Call Home: State-Sanctioned Monopolization in the Prison Phone Industry." Prison Policy Initiative, 2012. https://static.prisonpolicy.org/phones/price_to_call_home.pdf.

Kuriyan, Renee and Kathy Kitner. "Constructing Class Boundaries: Gender, Aspirations, and Shared Computing." *Information Technologies & International Development* 5, no. 1 (2009): 17–29.

Lagesen, Vivian Anette. "A Cyberfeminist Utopia? Perceptions of Gender and Computer Science among Malaysian Women Computer Science Students and Faculty." *Science, Technology, & Human Values* 33, no. 1 (2008): 5–27.

Laroia, Asheesh. "Diversity in Free Software: South Asians As an Example." *Asheeshworld* blog, December 18, 2009. http://www.asheesh.org/note/debian/indians.html.

Lara, Juan De. *Inland Shift: Race, Space and Capital in Southern California*. Berkeley: University of California Press, 2018.

Latour, Bruno. "Give Me a Laboratory and I Will Raise the World." In *Science Observed: Perspectives on the Social Study of Science*, edited by Karin Knorr-Cetina and Michael Mulkay. London: Sage, 1983.

———. *Science in Action*. Cambridge, MA: Harvard University Press, 1987.

Leach, James, Dawn Nafus, and Bernhard Krieger. "Freedom Imagined: Morality and Aesthetics in Open Source Software Design." *Ethnos* 74, no. 1 (2009): 51–71.

Leavitt, Lynda. "DARPA's Maple Leaf Remote Control Drone Takes First Flight (video)." *Engadget*, August 11, 2011. https://www.engadget.com/2011/08/11/darpas-maple-leaf-remote-control-drone-takes-first-flight-vide/.

Lecher, Colin. "James Damore Sues Google for Allegedly Discriminating against Conservative White Men." *Verge*, January 8, 2018. https://www.theverge.com/2018/1/8/16863342/james-damore-google-lawsuit-diversity-memo.

Lerman, Nina, Ruth Oldenziel, and Arwen P. Mohun, eds. *Gender & Technology: A Reader*. Baltimore: Johns Hopkins University Press, 2003.

Levy, Steven. *Hackers: Heroes of the Computer Revolution*. New York: Delta Publishing, 1984.

Light, Jennifer S. "When Computers Were Women." *Technology and Culture* 40, no. 3 (1999): 455–83.

Lin, Yuwei. "Embodying Hacker Culture in Women-Friendly Free Software Groups." Presented to Codes and Conduct Workshop, ESRC National Centre for e-Social Science, University of Manchester. November 19–20, 2007, Lancaster, United Kingdom.

———. "Gender Dimensions of FLOSS Development." *Mute* 2, no. 1 (2005): 38–42.

———. "A Techno-Feminist View on the Free and Open Source Software Development." *Encyclopedia of Gender and Information Technology*. Hershey, PA: Idea Group, 2006a.

———. "Women in the Free/Libre Open Source Software Development." In *Encyclopedia of Gender and Information Technology*. Hershey, PA: Idea Group, 2006b.

Lindtner, Silvia. "Hacking with Chinese Characteristics: The Promises of the Maker Movement against China's Manufacturing Culture." *Science, Technology, & Human Values* 40, no. 5 (2015): 854–79.

———. *Prototype Nation: The Maker Movement and the Promise of Entrepreneurial Living in China.* Princeton, NJ: Princeton University Press, forthcoming.

———, Garnet Hertz, and Paul Dourish. "Emerging Sites of HCI Innovation: Hackerspaces, Hardware Startups & Incubators." *CHI Proceedings*, April 26–May 1, 2014, Toronto, ON, Canada.

LinuxChix website. Accessed December 19, 2009. http://www.linuxchix.org/.

Lovecruft, Isis Agora. "The Forest for the Trees." *Patterns in the Void* (blog). Last modified June 16, 2016. https://blog.patternsinthevoid.net/the-forest-for-the-trees.html.

Lupton, Deborah. *The Quantified Self.* Cambridge, UK: Polity Press, 2016.

Lyndgate, Anthony. "What the SpaceX Explosion Means for Elon Musk and Mark Zuckerberg." *New Yorker,* September 2, 2016. https://www.newyorker.com/tech/elements/what-the-spacex-explosion-means-for-elon-musk-and-mark-zuckerberg.

Macrina, Alison. Medium.com, June 16, 2016. https://medium.com/@flexlibris/theres-really-no-such-thing-as-the-voiceless-92b3fa45134d#.ykw0kpfcl.

Manivannan, Vyshali. "FCJ-158 Tits or GTFO: The Logics of Misogyny on 4chan's Random–/b." *The Fibreculture Journal* 22 (2013). http://twentytwo.fibreculturejournal.org/fcj-158-tits-or-gtfo-the-logics-of-misogyny-on-4chans-random-b/.

Marcus, George E. "Ethnography in/of the World System: The Emergence of Multi-Sited Ethnography." *Annual Review of Anthropology* 24, no. 1 (1995): 95–117.

Marechal, Nathalie. "Use Signal, Use Tor? The Political Economy of Digital Rights Technology." PhD dissertation, University of Southern California, 2018.

Margolis, Jane. *Stuck in the Shallow End: Education, Race, and Computing.* Cambridge, MA: MIT Press, 2010.

———, and Allan Fisher. *Unlocking the Clubhouse: Women in Computing.* Cambridge, MA: MIT Press, 2003.

Markell, Patchen. *Bound by Recognition.* Princeton, NJ: Princeton University Press, 2003.

Martin, Aryn, Natasha Myers, and Ana Viseu. "The Politics of Care in Technoscience." *Social Studies of Science* 45, no. 5 (2015): 625–41.

Martinez, Christopher/Cristóbal. "Tecno-Sovereignty: An Indigenous Theory and Praxis of Media Articulated through Art, Technology, and Learning." PhD dissertation, Arizona State University, 2015.

Marvin, Carolyn. *When Old Technologies Were New.* Oxford: Oxford University Press, 1988.

Marwick, Alice. "Donglegate: Why the Tech Community Hates Feminists." *Wired,* March 23, 2013. http://www.wired.com/2013/03/richards-affair-and-misogyny-in-tech/.

Marx, Leo. "Technology: The Emergence of a Hazardous Concept." *Technology and Culture* 51, no. 3 (2010): 561–77.

Mavhunga, Clapperton. *Transient Workspaces: Technologies of Everyday Innovation in Zimbabwe.* Cambridge, MA: MIT Press, 2014.

Maxigas. "Hacklabs and Hackerspaces—Tracing Two Genealogies." *Journal of Peer Production* 2 (2012). http://peerproduction.net/issues/issue-2/peer-reviewed-papers/hacklabs-and-hackerspaces/.

McCall, Leslie. *Complex Inequality: Gender, Class, and Race in the New Economy.* New York: Routledge, 2001.

McGranahan, Carole. "What Is Ethnography? Teaching Ethnographic Sensibilities without Fieldwork." *Teaching Anthropology* 4 (2014): 23–36.

McIlwain, Charlton. *Black Software*. Oxford: Oxford University Press, 2019.

McInerney, Paul-Brian. "Technology Movements and the Politics of Free/Open Source Software." *Science, Technology, & Human Values* 34, no. 2 (2009): 206–33.

McPherson, Tara. "U.S. Operating Systems at Mid-Century: The Intertwining of Race and UNIX." In *Race after the Internet*, edited by Lisa Nakamura and Peter Chow-White. New York: Routledge, 2011.

McRobbie, Angela. *The Aftermath of Feminism: Gender, Culture, and Social Change*. London: Sage, 2008.

———. *Be Creative: Making a Living in the New Culture Industries*. Cambridge, UK: Polity Press, 2016.

———. "Clubs to Companies: Notes on the Decline of Political Culture in Speeded Up Creative Worlds." *Cultural Studies* 16, no. 4 (2002): 516–31.

Medina, Eden. *Cybernetic Revolutionaries: Technology and Politics in Allende's Chile*. Cambridge, MA: MIT Press, 2011.

———, Ivan da Costa Marques, and Christina Holmes. *Beyond Imported Magic: Essays on Science, Technology, and Society in Latin America*. Cambridge, MA: MIT Press, 2014.

Mellström, Ulf. "The Intersection of Gender, Race and Cultural Boundaries, or Why Is Computer Science in Malaysia Dominated by Women?" *Social Studies of Science* 39, no. 6 (2009): 885–907.

Metzlar, Donna. "Interview with Donna Metzlar (GenderChangers) by Gabriella Coleman," Institute of Network Cultures. Last modified March 7, 2009. https://vimeo.com/4090016.

Milan, Stefania. *Social Movements and Their Technologies: Wiring Social Change*. New York: Palgrave Macmillan, 2013.

Milberry, Kate. "(Re)making the Internet: Free Software and the Social Factory Hack." In *DIY Citizenship: Critical Making and Social Media*, edited by Matt Ratto and Megan Boler. Cambridge, MA: MIT Press, 2014.

Miller, Daniel, and Don Slater. *The Internet: An Ethnographic Approach*. Oxford: Berg, 2000.

Miltner, Kate M. "Anyone Can Code? The Coding Fetish and the Politics of Sociotechnical Belonging." PhD dissertation, University of Southern California, 2019.

Misa, Thomas, ed. *Gender Codes*. Hoboken, NJ: Wiley & Sons, Inc./IEEE Computer Society, 2010.

Mislán, Cristina, and Amalia Dache-Gerbino. "*Not* a Twitter Revolution: Anti-neoliberal and Antiracist Resistance in the Ferguson Movement." *International Journal of Communication* 12 (2018): 2622–40.

Mohanty, Chandra. *Feminism without Borders: Decolonizing Theory, Practicing Solidarity*. Durham, NC: Duke University Press, 2003.

———. "Under Western Eyes: Feminist Scholarship and Colonial Discourses." *Boundary* 2 (1984): 333–58.

Mr-hank. "Hi, I'm the Guy Who Made a Comment about Big Dongles . . ." *Hacker News*, March 2013. https://news.ycombinator.com/item?id=5398681.

Myers, Natasha. *Rendering Life Molecular: Models, Modelers, and Excitable Matter*. Durham, NC: Duke University Press, 2015.

Myrpl. "Github Introduces 'Code of Conduct' that 'Promotes Equality', Programmers React." *The Red Pill Network*, 2015. https://www.forums.red/p/TheRedPill/3946/github_introduces_code_of_conduct_that_promotes_equality_pro.

Nafus, Dawn. "'Patches Don't Have Gender': What Is Not Open in Open Source Software." *New Media & Society* 14, no. 4 (2012): 669–83.

———, James Leach, and Bernhard Krieger. "Free/Libre and Open Source Software: Policy Support (FLOSSPOLS), Gender: Integrated Report of Findings." University of Cambridge, 2006.

———, and Gina Neff. *Self-Tracking*. Cambridge, MA: MIT Press, 2016.

Nakamura, Lisa. *Digitizing Race*. Minneapolis: University of Minnesota Press, 2007.

———. "Glitch Racism." *Culture Digitally* (blog). Last modified December 10, 2013. http://culturedigitally.org/2013/12/glitch-racism-networks-as-actors-within-vernacular-internet-theory/.

———. "Indigenous Circuits: Navajo Women and the Racialization of Early Electronic Manufacture." *American Quarterly* 66 no. 4 (2014): 919–41.

Nakamura, Lisa, and Peter Chow-White. *Race after the Internet*. New York: Routledge, 2011.

National Center for Women & Information Technology. n.d. Accessed November 22, 2018. https://www.ncwit.org.

Nelsen, R. Arvid. "Race and Computing: The Problem of Sources, the Potential of Prosopography, and the Lesson of *Ebony* Magazine." *IEEE Annals of the History of Computing* 39, no. 1 (2017): 29–51.

Nelson, Alondra. *Body and Soul: The Black Panther Party and the Fight against Medical Discrimination*. Minneapolis: University of Minnesota Press, 2013.

———. "Introduction: Future Texts." *Social Text* 20, no. 2 (2002): 1–15.

Newitz, Annalee, and Charles Anders, eds. *She's Such a Geek: Women Write about Science, Technology and Other Nerdy Stuff*. Emeryville, CA: Seal Press, 2006.

Nguyen, Lilly U. "Infrastructural Action in Vietnam: Inverting the Techno-Politics of Hacking in the Global South." *New Media & Society* 18, no. 4 (2016): 637–52.

Noble, Safiya, *Algorithms of Oppression*. New York: New York University Press, 2018.

———, and Brendesha M. Tynes. *The Intersectional Internet: Race, Sex, Class, and Culture Online*. New York: Peter Lang International Academic Publishers, 2016.

Ohlheiser, Abby. "The Woman behind 'Me Too' Knew the Power of the Phrase When She Created It—10 Years Ago." *Washington Post*, October 19, 2017. https://www.washingtonpost.com/news/the-intersect/wp/2017/10/19/the-woman-behind-me-too-knew-the-power-of-the-phrase-when-she-created-it-10-years-ago/.

Oldenziel, Ruth. "Boys and Their Toys: The Fisher Body Craftsman's Guild, 1930–1968, and the Making of a Male Technical Domain." *Technology and Culture* 38, no. 1 (1997): 60–96.

———. *Making Technology Masculine. Men, Women, and Modern Machines in America, 1870–1945*. Amsterdam: Amsterdam University Press, 1999.

O'Leary, Amy. "Worries over Defense Department Money for 'Hackerspaces.'" *New York Times*, October 5, 2012. http://www.nytimes.com/2012/10/06/us/worries-over-defense-dept-money-for-hackerspaces.html.

Orr, Julian. "Sharing Knowledge, Celebrating Identity: Community Memory in a Service Culture." In *Collective Remembering*, edited by David Middleton and Derek Edwards. London: Sage, 1990.

Osman, Wazhmah. "Jamming the Simulacrum: On Drones, Virtual Reality, and Real Wars." In *Culture Jamming: Activism and the Art of Cultural Resistance*, edited by Marilyn DeLaure and Moritz Fink. New York: New York University Press, 2017.

Paolucci, Denise. "Overcoming Impostor Syndrome." Ada Initiative, posted on YouTube, June 26, 2013. https://www.youtube.com/watch?v=zZg9rax-ky4.

Parks, Lisa. "Drones, Vertical Mediation, and the Targeted Class." *Feminist Studies* 42, no. 1 (2016): 227–35.

Patterson, Meredith. "When Nerds Collide: My Intersectionality Will Have Weirdos or It Will Be Bullshit." Medium.com, March 23, 2014. https://medium.com/@maradydd/when-nerds-collide-31895b01e68c.

Paulitz, Tanja, Susanne Kink, and Bianca Prietl. "Into the Wild? Effects of Gender Equality Politics on Gender Studies Ethnographic Field Work in Engineering and Scientific Educational Cultures." Presentation to Annual Meeting of the Society for Social Studies of Science (4S), Copenhagen, Denmark, October 17–20, 2012.

Peña, Carolyn de la. "The History of Technology, the Resistance of Archives, and the Whiteness of Race." *Technology and Culture* 51, no. 4 (2010): 919–37.

Penny, Laurie. "How to Be a Genderqueer Feminist." *Buzzfeed*, October 31, 2015. https://www.buzzfeednews.com/article/lauriepenny/how-to-be-a-genderqueer-feminist#.vsxy7PXLQ.

Peppler, Kylie, Melissa Gresalfi, Katie Salen Tekinbas, Rafi Santo, and Leah Buechley. *Soft Circuits: Crafting e-Fashion with DIY Electronics*. Cambridge, MA, MIT Press, 2014.

Philip, Kavita, Lilly Irani, and Paul Dourish. "Postcolonial Computing: A Tactical Survey." *Science, Technology, & Human Values* 37, no. 1 (January 2012): 3–29.

Phillips, Whitney. *This Is Why We Can't Have Nice Things: Mapping the Relationship between Online Trolling and Mainstream Culture*. Cambridge, MA: MIT Press, 2015.

Piepmeier, Alison. *Girl Zines: Making Media, Doing Feminism*. New York: New York University Press, 2009.

Pinch, Trevor, and Frank Trocco. *Analog Days: The Invention and Impact of the Moog Synthesizer*. Cambridge, MA: Harvard University Press, 2002.

PlayHaven. "Addressing Pycon." Last modified March 21, 2013. https://web.archive.org/web/20130326120225/http://blog.playhaven.com/addressing-pycon/.

Polletta, Francesca. *Freedom Is an Endless Meeting: Democracy in American Social Movements*. Chicago: University of Chicago Press, 2002.

Portwood-Stacer, Laura. *Lifestyle Politics and Radical Activism*. New York: Continuum Press, 2013.

Poster, Winifred. "Cybersecurity Needs Women." *Nature* 555 (2018): 577–80.

———. "Filtering Diversity: A Global Corporation Struggles with Race, Class, and Gender in Employment Policy." *American Behavioral Scientist* 52 no. 3 (2008): 307–41.

———. "Who's on the Line? Indian Call Center Agents Pose as Americans for U.S.-Outsourced Firms." *Industrial Relations* 46, no. 2 (2007): 271–304.

———, and George Wilson. "Introduction: Race, Class, and Gender in Transnational Labor Inequality." *American Behavioral Scientist* 52, no. 3 (2008): 295–305.

———, and Zakia Salime. "Limits of Microcredit: Transnational Feminism and USAID Activities in the United States and Morocco." In *Women's Activism and Globalization: Linking Local Struggles and Transnational Politics*, edited by Nancy A. Naples and Manisha Desai. New York: Routledge, 2002.

Postigo, Hector. "From *Pong* to *Planet Quake*: Post-Industrial Transitions from Leisure to Work." *Information, Communication & Society* 6, no. 4 (2003): 593–607.

Powell, Alison. "Argument-By-Technology: How Technical Activism Contributes to Internet Governance." In *Research Handbook on Governance of the Internet*, edited by Ian Brown. Cheltenham, UK: Edward Elgar, 2013: 198–220.

———. "Democratizing Production through Open Source Knowledge: From Open Software to Open Hardware." *Media, Culture & Society* 34, no. 6 (2012): 691–708.

Prasad, Pritha. "Beyond Rights as Recognition: Black Twitter and Posthuman Coalitional Possibilities." *Prose Studies: History, Theory, Criticism* 38, no. 1 (2016): 50–73.

Prasad, Sandeep. "Is This Really What a Feminist Government Looks Like?" *Huffington Post*, March 7, 2018. https://www.huffingtonpost.ca/sandeep-prasad/canadian-government-feminist-policies_a_23374762/.

Puente, Sonia N. "From Cyberfeminism to Technofeminism: From an Essentialist Perspective to Social Cyberfeminism in Certain Feminist Practices in Spain." *Women's Studies International Forum* 31, no. 6 (2008): 434–40.

Puig de la Bellacasa, Maria. "Matters of Care in Technoscience: Assembling Neglected Things." *Social Studies of Science* 41, no. 1 (2011): 85–106.

Pullin, Graham. *Design Meets Disability*. Cambridge, MA: MIT Press, 2011.

Pursell, Carroll. "The Rise and Fall of the Appropriate Technology Movement in the United States, 1965–1985." *Technology and Culture* 34, no. 3 (1993): 629–37.

PyCon US. Code of Conduct. 2013. Accessed July 27, 2016. https://us.pycon.org/2013/about/code-of-conduct/.

Qiu, Jack. *Goodbye iSlave: A Manifesto for Digital Abolition.* Urbana: University of Illinois Press, 2016.

Rabinbach, Anson. *The Human Motor*. Berkeley: University of California Press, 1992.

Radway, Janice. "The Body Project of Girl Zines." *International Journal of Communication* 4 (2010): 224–25.

Raja, Tasneem. "Is Coding the New Literacy?" *Mother Jones*, July/August 2014. http://www.motherjones.com/media/2014/06/computer-science-programming-code-diversity-sexism-education.

Rankin, Joy. *A People's History of Computing in the United States.* Cambridge, MA: Harvard University Press, 2018.

Ratto, Matt. "Critical Making: Conceptual and Material Studies in Technology and Social Life." *The Information Society* 27, no. 4 (2011): 252–60.

Raymond, Eric. "The Cathedral and the Bazaar." *First Monday* 3, no. 3 (1998). http://firstmonday.org/htbin/cgiwrap/bin/ojs/index.php/fm/article/view/578/499.

———. *The New Hacker's Dictionary*, 3rd ed. Cambridge, MA: MIT Press, 1996.

Reagle, Joseph. "'Free As in Sexist?': Free Culture and the Gender Gap," *First Monday* 18, no. 1 (2013).

———. "Naïve Meritocracy and the Meanings of Myth." *Ada: A Journal of Gender & New Media Technology* 11 (2017). http://adanewmedia.org/2017/05/issue11-reagle/.

———. "The Obligation to Know: From FAQ to Feminism 101." *New Media & Society* 18, no. 5 (2016): 691–707.

Rentschler, Carrie. "Rape Culture and the Feminist Politics of Social Media." *Girlhood Studies* 7, no. 1 (2014): 65–82.

———, and Samantha Thrift. "Doing Feminism in the Network: Networked Laughter and the 'Binders Full of Women' Meme." *Feminist Theory* 16, no. 3 (2015): 329–59.

Renzi, Alessandra. "Info-Capitalism and Resistance: How Information Shapes Social Movements." *Interface: A Journal for and about Social Movements* 7, no. 2 (2015): 98–119.

Rickford, Russell. "Black Lives Matter: Toward a Modern Practice of Mass Struggle." *New Labor Forum* 25, no. 1 (2016): 34–42.

Riseup.net. Accessed January 1, 2010. http://riseup.net/.

Roberts, Tony. "How Tech Geeks in Africa are Transforming IT Education." Computer World (April 2012). http://www.computerweekly.com/opinion/How-tech-geeks-in-Africa-are-transforming-IT-education.

Rodino-Colocino, Michelle. "Geek Jeremiads: Speaking the Crisis of Job Loss by Opposing Offshored and H-1B Labor." *Communication and Critical/Cultural Studies* 9, no. 1 (2012): 22–46.

Rosenzweig, Roy. "Can History Be Open Source? Wikipedia and the Future of the Past." *Journal of American History* 93, no. 1 (2006): 117–46.

———. "Wizards, Bureaucrats, Warriors, and Hackers: Writing the History of the Internet." *American Historical Review* 103, no. 5 (1998): 1530–52.

Rosner, Daniela. *Critical Fabulations*. Cambridge, MA: MIT Press, 2018.

———, and Sarah E. Fox. "Legacies of Craft and the Centrality of Failure in a Mother-Operated Hackerspace." *New Media & Society* 18, no. 4 (2016): 558–80.

S.S.L. Nagbot. "Feminist Hacking/Making: Exploring New Gender Horizons of Possibility." *Journal of Peer Production* 8 (2016): 1–10. http://peerproduction.net/issues/issue-8-feminism-and-unhacking/feminist-hackingmaking-exploring-new-gender-horizons-of-possibility/.

Said, Edward W. "Representing the Colonized: Anthropology's Interlocutors." *Critical Inquiry* 15, no. 2 (1989): 205–25.

Salehi, Niloufar, and Tech Workers Coalition. "To the Feminists at Grace Hopper." *Medium.com*, October 2, 2017. https://medium.com/tech-workers-coalition/to-the-feminists-at-grace-hopper-edbbb43f7e5d.

———, and Michael S. Bernstein. "Hive: Collective Design through Network Rotation." Proceedings: ACM Human-Computer Interaction 2, CSCW, Article 151 (November 2018). https://doi.org/10.1145/3274420.

Sandoval, Chela. *Methodology of the Oppressed*. Minneapolis: University of Minnesota Press, 2000.

Schiebinger, Londa. *Gendered Innovations in Science and Engineering*. Stanford, CA: Stanford University Press, 2008.

Schiller, J. Zach. "On Becoming the Media: Low Power FM and the Alternative Public Sphere." In *Media and Public Spheres*, edited by Richard Butsch. New York: Palgrave Macmillan, 2007.

Scholz, Trebor, and Nathan Schneider, eds. *Ours to Hack and to Own: The Rise of Platform Cooperativism, a New Vision for the Future of Work and a Fairer Internet*. New York: OR Books, 2017.

Schulte, Stephanie. *Cached: Decoding the Internet in Global Popular Culture*. New York: New York University Press, 2013.

Scott, Joan W. "Gender: A Useful Category of Historical Analysis." *American Historical Review* 91, no. 5 (1986): 1053–75.

Sender, Katherine. *Business Not Politics: The Making of the Gay Market*. New York: Columbia University Press, 2005.

Shaw, Adrienne. "The Internet Is Full of Jerks, Because the World Is Full of Jerks: What Feminist Theory Teaches Us about the Internet." *Communication and Critical/Cultural Studies* 11, no. 3 (2014): 273–77.

Shaw, Aaron, and Benjamin Mako Hill. "Laboratories of Oligarchy? How the Iron Law Extends to Peer Production." *Journal of Communication* 64, no. 2 (2014): 215–38.

Sims, Christo. *Disruptive Fixation: School Reform and the Pitfalls of Techno-Idealism*. Princeton, NJ: Princeton University Press, 2017.

Singh, Rianka. "Platform Feminism: Protest and the Politics of Spatial Organization." *Ada: A Journal of Gender, New Media, and Technology* 14 (2018). https://adanewmedia.org/2018/11/issue14-singh/.

Sivek, Susan Currie. "'We Need a Showing of All Hands': Technological Utopianism in *MAKE* Magazine." *Journal of Communication Inquiry* 35, no. 3 (2011): 187–209.

Skud. "Questioning the merit of meritocracy," *Geek Feminism*, November 29, 2009. http://geekfeminism.org/2009/11/29/questioning-the-merit-of-meritocracy/.

———. "Why I'm Not an Open Source Person Any More." Infotropism.com, January 28, 2011. http://infotrope.net/2011/01/28/why-im-not-an-open-source-person/.

Slaton, Amy. "Exit, Stage Left: Towards Transformative Critiques of Diversity," November 28, 2017. https://amyeslaton.com/exit-stage-left-towards-transformative-critiques-of-diversity/.

———. *Race, Rigor, and Selectivity in U.S. Engineering: The History of an Occupational Color Line*. Cambridge, MA: Harvard University Press, 2010.

Smith, Mark, and Denise Paolucci. "Build Your Own Contributors, One Part at a Time" slideshow, n.d.

Söderberg, Johan. *Hacking Capitalism*. New York: Routledge, 2008.

———, and Alessandro Delfanti. "Hacking Hacked! The Life Cycles of Digital Innovation." *Science, Technology, & Human Values* 40, no. 5 (2015): 793–98.

Sørensen, Knut, Wendy Faulkner, and Els Rommes. *Technologies of Inclusion: Gender in the Information Society*. Trondheim: Tapir Akademisk Forlag, 2011.

Spertus, Ellen. "Why Are There So Few Female Computer Scientists?" *MIT Artificial Intelligence Laboratory Technical Report-1315*, 1991.

Star, Susan Leigh. "The Ethnography of Infrastructure." *American Behavioral Scientist* 43, no. 3 (1999): 377–91.

———. "Power, Technology, and the Phenomenology of Conventions: On Being Allergic to Onions." In *A Sociology of Monsters: Essays on Power, Technology, and Domination*, edited by John Law. New York: Routledge, 1991.

———, and Griesemer, James. " 'Translations' and Boundary Objects: Amateurs and Professionals in Berkeley's Museum of Vertebrate Zoology, 1907–39." *Social Studies of Science* 19, no. 3 (1989): 387–420.

Stein, Arlene. "Sex, Truths, and Audiotape: Anonymity and the Ethics of Exposure in Public Ethnography." *Journal of Contemporary Ethnography* 39, no. 5 (2010): 554–68.

Sterne, Jonathan. "Out with the Trash: On the Future of New Media." In *Residual Media*, edited by Charles Acland, 16–31. Minneapolis: University of Minnesota Press, 2007.

———. "There Is No Music Industry." *Media Industries Journal* 1, no. 1 (2014): 50–55.

Stoller, Matt. "Why We Need to Break Up Amazon—And How to Do It." *Naked Capitalism*, October 17, 2014. http://www.nakedcapitalism.com/2014/10/matt-stoller-need-break-amazon.html.

Stone, Sandy. "The Empire Strikes Back: A Post-Transsexual Manifesto." *Camera Obscura: Feminism, Culture, and Media Studies* 10, no.2 (1992): 150–76. https://doi.org/10.1215/02705346-10-2_29-150.

Strathern, Marilyn. "Robust Knowledge and Fragile Futures." In *Global Assemblages: Technology, Politics, and Ethics as Anthropological Problems*, edited by Aihwa Ong and Stephen J. Collier. Malden, MA: Blackwell Publishing, 2005.

Strauss, Anselm. "A Social World Perspective." *Studies in Symbolic Interaction* 1, no. 1 (1978): 119–28.

Streeter, Thomas. "The Internet as a Structure of Feeling: 1992–1996." *Internet Histories* 1, no. 1–2 (2017): 79–89.

———. *The Net Effect: Technology, Romanticism, Capitalism*. New York: New York University Press, 2010.

———. *Selling the Air: A Critique of the Policy of Commercial Broadcasting in the United States*. Chicago: University of Chicago Press, 1996.

Sturken, Marita, and Douglas Thomas. "Introduction." In *Technological Visions: The Hopes and Fears That Shape New Technologies*, edited by Marita Sturken, Douglas Thomas, and Sandra Ball-Rokeach. Philadelphia: Temple University Press, 2004.

Suchman, Lucy. "Feminist STS and the Sciences of the Artificial." In *New Handbook of Science, Technology and Society*, edited by E. Hackett, O. Amsterdamska, M. Lynch, and J. Wajcman, 139–64. Cambridge, MA: MIT Press, 2008.

———. "Located Accountabilities in Technology Production," Centre for Science Studies, Lancaster University, Lancaster, UK, 2003.

Syster Server. Accessed September 25, 2016. https://systerserver.net/.

Takahashi, Dean. "Open Source Model Disrupts the Commercial Drone Business." *VentureBeat*, July 27, 2012. https://venturebeat.com/2012/07/27/open-source-model-disrupts-the-commercial-drone-business/.

Takhteyev, Yuri. *Coding Places: Software Practice in a South American City*. Cambridge, MA: MIT Press, 2012.

Tal, Kali. "'Duppies in the Machine': White Cyberculture Critics Read Race." Presented to the American Studies Association Annual Conference, Detroit, MI, October, 2000.

TallBear, Kim. "An Indigenous Reflection on Working beyond the Human/Not Human." *GLQ: A Journal of Lesbian and Gay Studies* 21, no. 2–3 (2015): 230–35.

Tarnoff, Ben. "Tech Workers Versus the Pentagon: An Interview with Kim." *Jacobin*, June 6, 2018. https://jacobinmag.com/2018/06/google-project-maven-military-tech-workers.

———. "Trump's Tech Opposition: An Interview with Kristen Sheets and Matt Schaefer." *Jacobin*, May 2, 2017. https://www.jacobinmag.com/2017/05/tech-workers-silicon-valley-trump-resistance-startups-unions.

Taylor, Charles. *Modern Social Imaginaries*. Durham, NC: Duke University Press, 2004.

Taylor, Craig. "Electric Ladyland." *Harper's Magazine*, April 2012.

Terrell J., A. Kofink, J. Middleton, C. Rainear, E. Murphy-Hill, C. Parnin, and J. Stallings. "Gender Differences and Bias in Open Source: Pull Request Acceptance of Women versus Men." *PeerJ Preprints*. July 26, 2016. https://doi.org/10.7287/peerj.preprints.1733v2.

Tkacz, Nathaniel. *Wikipedia and the Politics of Openness*. Chicago: University of Chicago Press, 2014.

Toombs, Austin, Shaowen Bardzell, and Jeffrey Bardzell. "The Proper Care and Feeding of Hackerspaces: Care Ethics and Cultures of Making." Proceedings of the 33rd Annual ACM Conference on Human Factors in Computing Systems, Seoul, April 18–23. New York: ACM Press, 2015: 629–38.

Tor Project (blog). "Solidarity against Online Harassment." Last modified December 11, 2014. https://blog.torproject.org/blog/solidarity-against-online-harassment.

Toupin, Sophie. "Feminist Hackerspaces as Safer Spaces?" *Feminist Journal of Art and Digital Culture* 27 (2015). http://dpi.studioxx.org/en/feminist-hackerspaces-safer-spaces.

———. "Feminist Hackerspaces: The Synthesis of Feminist and Hacker Cultures." *Journal of Peer Production* 5 (2016a). http://peerproduction.net/issues/issue-5-shared-machine-shops/peer-reviewed-articles/feminist-hackerspaces-the-synthesis-of-feminist-and-hacker-cultures/.

———. "Gesturing towards 'Anti-Colonial' Hacking and Its Infrastructure." *Journal of Peer Production* 9 (2016b). http://peerproduction.net/editsuite/issues/issue-9-alternative-internets/peer-reviewed-papers/anti-colonial-hacking/.

Tronto Joan. *Moral Boundaries: A Political Argument for an Ethic of Care*. New York: Routledge, 1993.

Tu, Thuy Linh, and Alondra Nelson. *Technicolor: Race, Technology, and Everyday Life*. New York: New York University Press, 2001.

Turkle, Sherry. *Life on the Screen*. New York: Simon & Schuster, 1995.

———. *The Second Self: Computers and the Human Spirit*. New York: Simon & Schuster, 1984.

Turner, Fred. "Burning Man at Google: A Cultural Infrastructure for New Media Production." *New Media & Society* 11, no. 1–2 (2009): 73–94.

———. *From Counterculture to Cyberculture*. Chicago: University of Chicago Press, 2006.

United States. "President Trump Signs Memorandum for STEM Education Funding." September 25, 2017. https://www.whitehouse.gov/articles/president-trump-signs-memorandum-stem-education-funding/.

United States. "Women in STEM." Whitehouse Office of Science & Technology Policy. n.d. Accessed February 17, 2015. http://www.whitehouse.gov/administration/eop/ostp/women.

Upadhya, R.K., and Tech Workers Coalition. "Tech Workers against Imperialism." November 21, 2018. https://medium.com/tech-workers-coalition/tech-workers-against-imperialism-2d8024e461a7.

Vowel, Chelsea. Indigenous Writes. Winnipeg: Highwater Press, 2016.

Wajcman, Judy. *Feminism Confronts Technology*. University Park: Pennsylvania State University Press, 1991.

———. "From Women and Technology to Gendered Technoscience." *Information, Community and Society* 10, no. 3 (2007): 287–98.

———. *Pressed for Time: The Acceleration of Life in Digital Capitalism*. Chicago: University of Chicago Press, 2015.

———. *TechnoFeminism*. Cambridge, UK: Polity, 2004.

Wakabayashi, Daisuke, Erin Griffith, Amie Tsang, and Kate Conger. "Google Walkout: Employees Stage Protest over Handling of Sexual Harassment." *New York Times*, November 1, 2018. https://www.nytimes.com/2018/11/01/technology/google-walkout-sexual-harassment.html.

Waksman, Steve. "California Noise: Tinkering with Hardcore and Heavy Metal in Southern California." *Social Studies of Science* 34, no. 5 (2004): 675–702.

Wallach, Hanna. "Women in Free/Open Source Software Development." 2nd Annual *Journal of Information Technology & Politics* Thematic Conference, Amherst, MA, May 6–7, 2010.

Warner, Michael. "Publics and Counterpublics." *Public Culture* 14, no. 1 (2002): 49–90.

———. *The Trouble with Normal: Sex, Politics, and the Ethics of Queer Life*. Cambridge, MA: Harvard University Press, 1999.

West, Sarah Myers. "Cryptographic Imaginaries and Networked Publics: A Cultural History of Encryption Technologies, 1967–2017." PhD dissertation, University of Southern California, 2018.

Willis, Paul. *Learning to Labor*. New York: Columbia University Press, 1981.

Winner, Langdon. *The Whale and the Reactor*. Chicago: University of Chicago Press, 1986.

Winter, Debra and Chuck Huff. "Adapting the Internet: Comments from a Women-Only Electronic Forum." *American Sociologist* 27, no. 1 (1996): 30–54.

Wired Magazine. "Your Own Personal Internet." *Wired Magazine*, June 30, 2006. http://www.wired.com/threatlevel/2006/06/your_own_person/.

Wisnioski, Matthew. *Engineers for Change*. Cambridge, MA: MIT Press, 2012.

Wolfson, Todd. *Digital Rebellion: The Birth of the Cyber Left*. Urbana: University of Illinois Press, 2014.

———. "From the Zapatistas to Indymedia: Dialectics and Orthodoxy in Contemporary Social Movements." *Communication, Culture & Critique* 5, no. 2 (2012): 149–70.

Wynne, Brian. "Misunderstood Misunderstandings: Social Identities and the Public Uptake of Science." In *Misunderstanding Science? The Public Reconstruction of Science and Technology*, edited by Alan Irwin and Brian Wynne. Cambridge: Cambridge University Press, 1996. 19–46.

Young, Angelo. "A Labor Movement Is Brewing within the Tech Industry." *Salon.com*, June 10, 2017. https://www.salon.com/2017/06/10/a-labor-movement-is-brewing-within-the-tech-industry/.

Young, Michael. "Down with Meritocracy." *Guardian*, June 29, 2001. http://www.theguardian.com/politics/2001/jun/29/comment.

Zuckerberg, Mark. "Building Global Community," February 16, 2017. https://www.facebook.com/notes/mark-zuckerberg/building-global-community/10103508221158471/.

INDEX

Note: Page numbers in italics indicate figures.

A NOTE ON THE TYPE

This book has been composed in Adobe Text and Gotham. Adobe Text, designed by Robert Slimbach for Adobe, bridges the gap between fifteenth- and sixteenth-century calligraphic and eighteenth-century Modern styles. Gotham, inspired by New York street signs, was designed by Tobias Frere-Jones for Hoefler & Co.